Urban Geography

Urban Geography

Ian Diego

MURPHY & MOORE
www.murphy-moorepublishing.com

Urban Geography
Ian Diego
ISBN: 978-1-63987-561-0 (Hardback)

MURPHY & MOORE
Published by Murphy & Moore Publishing,
1 Rockefeller Plaza,
New York City, NY 10020, USA

Cataloging-in-Publication Data

Urban geography / Ian Diego.
p. cm.
Includes bibliographical references and index.
ISBN 978-1-63987-561-0
1. Urban geography. 2. Geography. I. Diego, Ian.
GF125 .U73 2022
307.76--dc23

For more information regarding Murphy & Moore Publishing and its products, please visit the publisher's website www.murphy-moorepublishing.com

Table of Contents

Preface

The branch of geography concerned with cities, urban life, built environment and urban processes is known as urban geography. It attempts to understand the flow of natural and economic resources, development of infrastructure, governance, institutional activities and socio-spatial inclusions and exclusions. Some of the major areas of study under this discipline are urban planning, urban anthropology and urban sociology. On the basis of their roles in the city system, urban geography classifies cities into central places, specialized-function cities and transportation cities. Central places act as service centers for local hinterlands. Functions such as mining, manufacturing, etc. are undertaken by specialized function cities. This book presents the complex subject of urban geography in the most comprehensible and easy to understand language. It elucidates new techniques and their applications in a multidisciplinary approach. This book will provide comprehensive knowledge to the readers.

A short introduction to every chapter is written below to provide an overview of thecontent of the book:

Chapter 1 - The area of study within geography which focuses on cities and urban processes is termed as urban geography. A few of the major areas of study within this field are urban morphology and urban poverty. This chapter has been carefully written to provide an easy introduction to these facets of urban geography; **Chapter 2** - The rate at which the population of an urban area increases is termed as urban growth. Some of the major issues connected with urban growth are overpopulation, unemployment, sanitation problems and shortage of water. This chapter discusses in detail these problems related to urban growth as well as the stages of urban growth; **Chapter 3** - The arrangement of land use in urban areas is known as urban structure. The process which is involved in developing and designing built environment and land use in urban areas is called urban planning. The topics elaborated in this chapter will help in gaining a better perspective about the different aspects of urban structure, planning and design; **Chapter 4** - The process of a shift in the population from rural areas to urban areas is called urbanization. Some of the concepts studied in relation to urbanization are suburbanization, overurbanization, counterurbanization and peri-urbanisation. All these diverse processes related to urbanization have been carefully analyzed in this chapter; **Chapter 5** - The relationship between the terrain and the composition, migration, distribution and growth of populations is studied under the discipline of population geography. Some of the major topics studied within population geography are demography, mortality rate and population policy. This chapter discusses in detail these aspects of geographical study of population; **Chapter 6** - The representation of demographic processes through a mathematical function or a set of functions connecting two or more measurable demographic variables is termed as a demographic model. It is used for the simplification of data. This chapter has been carefully written to provide an easy understanding of the varied facets of demographic models.

Finally, I would like to thank my fellow scholars who gave constructive feedback and my family members who supported me at every step.

Ian Diego

1

Introduction to Urban Geography

The area of study within geography which focuses on cities and urban processes is termed as urban geography. A few of the major areas of study within this field are urban morphology and urban poverty. This chapter has been carefully written to provide an easy introduction to these facets of urban geography.

Urban geography as a discipline evolved in the twentieth century. Over the time it has developed into a well-established discipline which deals with the study of urban settlements within the framework of their geographical setting. One can say that the scope of the sub-discipline constitutes the study of origin of urban settlements, their morphology and its development, their functions in and around their environs. With the increase of population and these settlements emerging as the magnets of economic, social and political developments; the discipline had gained importance in social sciences. The earlier urban geographers mainly concentrated on the physical aspects of the cities and their situation. The main emphasis was on the relationship that existed between the location and the structure of some particular cities and their surroundings.

With time the framework of urban geographers changed and in present times two common approaches are identifiable. The first one mainly deals with cities as distinct phenomenon located on the surface of the earth. They usually study the distribution, size, function and growth rate of urban settlements along with the spatial interactions between different urban centres. The second viewpoint analyses cities with respect to their morphology (layout and build up area) and intensity of land use within the city. Within this framework some writers have also started analysing problems related to urban growth and development.

Over the period of time the scope of urban geography has extended and now basically includes the areal association of activities within urban places; the economic base of cities; patterns of distribution of the cities over the earth surface; distribution of different geographical phenomenon within the city and the spatial interactions of one city with the other.

It can be said that urban geography has a distinctive focus with its primary concern being the study of association of activities in urban areas, association of land use and other features. The main focus is on the interpretation of patterns and relationships among different cities as well as that between urban and non-urban places.

Various definitions have been given by different urban geographers to understand the nature and scope of urban geography. Few have been listed here to understand how the scope and subject matter of urban geography has evolved over time.

Urban Geographers define urban geography as the study of a city directing the neighbouring region. The city behaves like a king in its hinterland. Urban geography is concerned with the study of the economic base of cities with interpretations of the associations between cities as man's habitat and economic activities within the cities and its hinterland.

As the geographer deals with the study of the uneven character of the earth's surface; and considerable proportion of population resides in urban settlements these settlements with its inhabitants and the buildings together are of special interest to the urban geographer. Moreover, the problems faced by the city dwellers are of utmost importance while studying urban geography.

The urban geographer always plays a dual role. Firstly, he examines cities in the context of location, characteristics, growth and their relationship with its hinterland and secondly, discusses city's morphology in terms of land use, social and cultural landscape, circulation patterns and also the components of the physical environment – all these in interrelation and interaction within the urban area.

This clearly states that that the process of evolution of urban geography is quite complex. Researchers have identified six elements that form the base of urban geography. These are:

- The Built environment.
- Human –environment relationship in an urban context.
- Social geography and social patterns in an urban context.
- City systems and functions: Macro scale.
- City systems and functions: Micro scale.
- Urban planning, policy and design.

Whatever the case may be, the written literature clearly states that urban geography in recent years has become more of a combination of various approaches and considerations of the city. It has become interdisciplinary in nature rather than being compartmentalised with in a structured geographical study.

Nature and Scope of Urban Geography

Aurousseau was among the first ones who gave an outline of the subject matter of the urban geography. He is of the view that since this part of geography embraces a large part of human geography if fails to be a specialized subject and therefore is not sure about the nature of urban geography. But after analysing various approaches he concludes that the regional study of towns and their functional study do form an important component of its scope. This gave impetus to the 'site and situation' and 'functional' approaches within this discipline.

The morphological approach gained momentum with the emergence of the Chicago School in the late 1920s. They paid attention to diverse social and economic factors that were responsible for the segregated land use in the city. Thus, now the scholars diverted their attention to the complexities of the cityscape rather than concentrating on the growth and layout of the cities. This gave foundation to the new urban geography where this discipline became more of an integrated systematic study. Urban geography is not about planning but is concerned with various factors which are inherent to the spatial and geographical structure of the city upon which planning should be based.

With planning gaining emphasis, functions too became important as now location was understood through functions that are what a town does or did in the past. The functions now also determine the pattern of city's growth and development. When urban geography started crystallising into a well define systematic study; came the quantitative revolution. Model building came into existence and theory had to be tested in reality; which usually involved statistical techniques; the most significant example being Christaller's central place theory.

Quantification although transformed vague descriptions into crisp models through which theory could be derived, it did not last long and collapsed. It became evident that explanations conceived at the initial level were not enough. The scholars moved towards what is called the behavioural approach. This approach was deep rooted therefore provided better and satisfactory answers. Now the studies were dominated with the studies of behaviour of the consumer and choices of the residence along with people's perception of the city and the opportunities it offers.

After the Second World War, the urban geographers acted more as consultants to different planning organisations and moved away from the "social conscience" approach which dealt with the study of spatial inequalities in urban settlements. Geographers identify three branches within the urban geography that were the result of these changes in the form of three different approaches. The first one is based on nomothetic philosophy and is quantitative in nature, where the geographer documents the spatial organization of the phenomenon. The second approach is behavioural in nature as it studies individual activities within their paraphernalia. The third approach is radical in nature as it stresses on inequality and constrains that the society imposes on the behaviour of certain identified groups within the city.

Over time urban geography has become less unified and so it has become very difficult to compartmentalise its scope. Nevertheless one can delineate the thrusts of the urban geography as shown in figure.

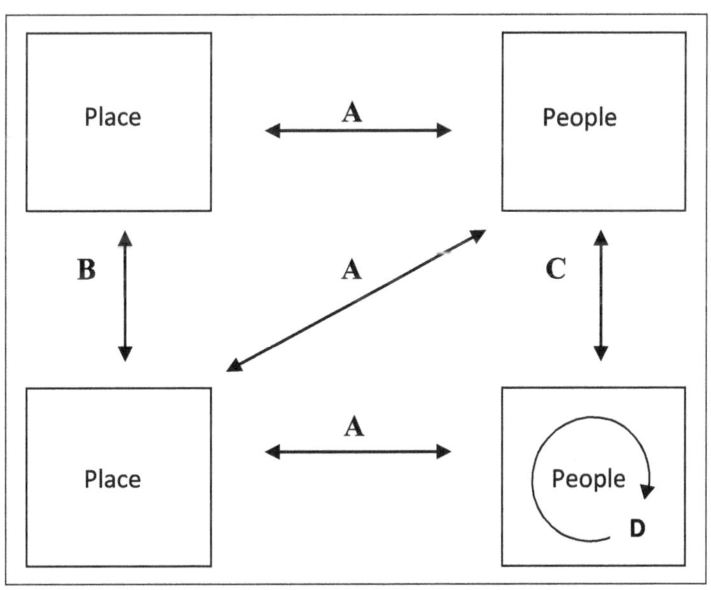

Thrusts of Study in Urban Geography.

Through this diagram he indicates the different inquiries that urban geography should look into. A involves relationship between a place and its people; B deals with association between different places; C is concerned with the study of relationship between people of different places and D is associated with the relations that exist within one place or among the people of one particular place.

A city is a geographical reservoir of roads and dwellings, a centre of commerce and administration, a type of society, even the cultural form of the mind of the urban people or the way of life style. Various attributes that can form the scope and content of this sub-discipline of geography have been summarized in the following diagram:

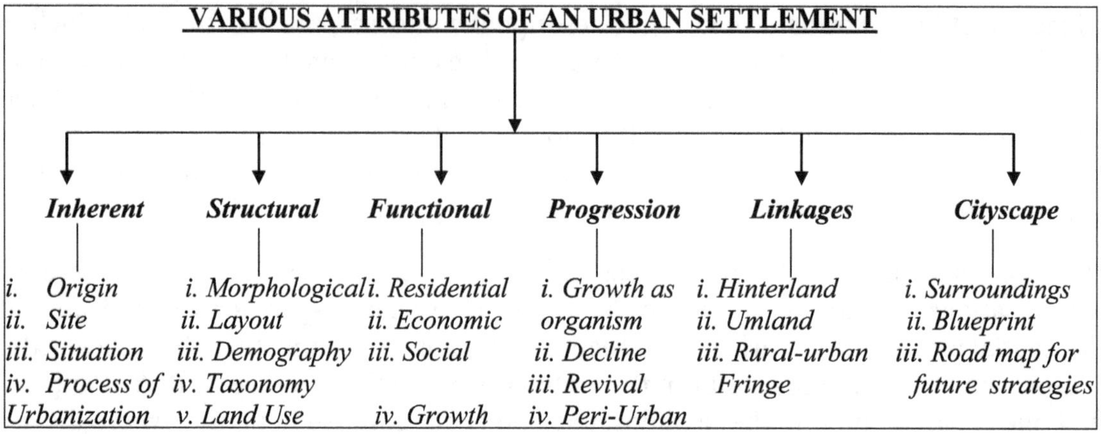

Meaning of Urban Place

The most important and immediate problem in deciding what 'urban' is? It is different from its equivalent, that is, the rural? In everyday life, we know that the difference between rural and urban is dependent on the function of that settlement that is in the former the main function is agriculture while in the latter the population by large is engaged in non-agricultural activities. In spite of this basic and predominant difference it is still difficult to define urban settlement as the process of urbanization is dynamic in nature. It becomes very difficult to confine the term in a water tight compartment. The work becomes more cumbersome as 'an urban place' is defined separately by different scholars and agencies. For example, a sociologist defines "urban" as a way of life or a condition of man that is characterised with certain attitudes like superficiality, ambiguity etc. But, for, geographers it is much more than that. They perceive urban in terms of location which has following features:

- The population density of such a settlement is relatively more than that of general population.
- Most of the dwellers are engaged in non-agricultural economic activities.
- The centre acts as a hub of social, cultural, administrative and economic activities.

An urban settlement can be regarded as a man-made habitat on the surface of the Earth; where economic activities are separated from the soil and are carried out in association with similar activities at a fixed location. City is a locale with a relatively large population living urbanism as their way of life.

Even the United Nations Demographic Year Book has given many examples, in which different countries that define demographic are included. The United Nations Organization defines permanent settlement with a minimum population of 20,000 in the form of an urban location. But in many countries, Botswana (5,000), Ethiopia (2,000), Argentina (2,000), Israel (2,000),

Czechoslovakia (5,000), Iceland (200), Norway (200), Portugal (10,000), Japan (50,000), Australia (1,000), India (5,000), etc.

But, the UN Demographic Year Book concludes: “There is no point in the continuum from large agglomerations to small clusters or scattered dwellings where urbanity disappears and rurality begins the division between urban and rural populations is necessarily arbitrary.” The review of these definitions reveals that the following characteristics can be identified to define a place in the form of urban. These are:

- A place designated by the administrative status.
- Minimum population.
- Minimum population density.
- A concept of reconciliation or loose dealing settlement to join or exclude suburban areas.
- Proportion of non-agricultural occupations.
- A functional character.

The census of 1981 has said the following places as urban:

- Municipal, City Board, Cantonment Board / Notified Town Area Centres.
- A minimum population of 5,000.
- 75 percent of men engaged in non-agricultural activities.
- A minimum population density per square km per person 400 or 1000 square meters per person.

Before accepting the meaning of urban and rural, two important facts should be kept in mind. One fact is that it is now impossible to identify the division line between rural and urban - two have been merged to spread in a way and a scenario has been presented that neither has been fully cultivated nor so in full tertiary activities.

Industrialization has been brought in large scale settlements which are not villages, but are the centre of settlement of agricultural population. Another problem is about the concept of urban, which is not static and changes over time with space. The proportion of population engaged in agricultural activities is the most effective measure. But in the modern times the proportion of population in rural areas is irreversible due to the arrival of capital and urban workers.

Thus, there is a point to conclude in this discussion that with the changing nature of both rural and urban, functional overlap has developed between the two, so the difference between urban and villagers have actually lost its meaning in reality. Therefore, one needs to know the properties of a Town; and to know this one has to understand the qualifications of a city. Many qualities of a city or city can be summarized as follows:

- Town is a kind of settlement, which has a much higher scope than a simple rural establishment.
- It does not represent the number of people in a vast area. But it represents a stage of civilization that is quite different from expressing rural lifestyle from a rural area.

- Their historical roots in cities and towns; Blache has mentioned that cities are especially near their origins (rituals, surnames, heroes, etc.) around the mythical halo.
- Towns and cities, along with the creatures of commerce and early events, politics like: Babylon, Athens, London, Paris, Delhi etc.

Basic Concept in Urban Geography

To further understand the nature of urban geography, which became quite complex and hybrid by the latter half of the Twentieth century certain basic concepts need to be explored.

Site-Situation Concept

With the passage of time, the natural situation of the city has been utilized by the resources available and the area's adaptability and surrounding area. Its growth and expansion is sometimes spread to spread the natural site so that it can be made beyond recognition. There was little possibility of the development of true urban geography in this context.

The objective was limited and it was not possible to explain complex economic work and social order. It was further emphasized that the application of the 'site and situation formula was meaningless "where there was no historical interest on the site, because the situation was seen in the context of the routes and not the streams of the movement". Such examples dismissed the stereotype of 'site and situation'.

Concept of Ecology

During the two world wars, the concept of plant ecology came into being and influenced geographical events. Due to increase in population in the city area, the ecological processes of the city have changed. Urban ecology affected relations with the surrounding areas of the city and it had an impact on people and their environment. Several studies emphasised on the opportunities and constrains posed by physical environment. Simultaneously the political and economic processes involved in the production of urban environmental infrastructures also became part of urban geography.

In recent years, the concept of urban ecosystem has become important component of this sub-discipline. Along with this the impact of urban societies on sustainable development has also been taken within its fold. These concerns have led to local scale initiatives.

Behaviouralism and Urban Centres

Researchers argument about the city and its development is related to the behaviour of its consumer, which is the perfect alternative to the use of land. It depends on three variables, these are:

- Value of residential unit - purchase cost or rent?
- Quality of residence,
- Relationship with work and place of neighbourhood?

Family income is an important component for the selection of site in the city, and it relies on the capabilities of its users to interact with space. As a general trend, it is clear that people with almost one-income groups make their choices for similar place.

Concept of Radicalism

An important aspect of the urban world is the concept of 'total change' brought into play by the development new infrastructure like metro; it has completely changed the idea of consumers. These reactions also influence urban planning and the capitalist cities were now bound to think on the rights of the people. Radicals believe that they provide the fundamental base on the lines of market inequality and thus provided alternative systems of 'socialist city' - the planning centre and the central direction for equality.

To conclude the discussion, urban geography is more than a mix of different approaches. This is probably near a systematic geographic study rather than multi-dimensional analysis of its purpose. Now urban geography has become less unified and so it becomes more difficult to present a neat structure. Whatever the argument may be the fundamental concern is spatial and that also with relation to the description and explanation of pattern is core and that makes it important even for urban geography like other sub disciplines.

Early Urban Settlements

Graeco-Roman Times

In early (post-Mycenaean) Greece the creation of new colonies and cities date from the 11th century BC. By the 8th century BC huts had been completely replaced by rectangular houses, and in Ionia in particular various cities were flourishing. Cities were founded as the administrative and defensive centres of new colonies, or by synoecism (the political amalgamation of existing communities,). Two major Greek contributions to city planning were the so-called gridiron rectangular system of street construction, and the zoning of cities into e.g. public, private, economic and sacred areas. Gridiron planning evolved in approximately the 9th century BC and was clearly evident in the 8th century city of Megara Hyblaea. The gridiron system is often erroneously called the "Hippodamian system" after a famous 5th century BC town planner who utilized the method. There was as yet no real coordination of city planning, but the evolving cities had certain features in common.

Most cities originated around a defensive position (usually an acropolis on high ground, with a palace and later temple) surrounded by a strong wall which followed the contours of the site, had forts at regular intervals and strong gates opposite the main streets. Defense potential, adequate water supply and economic potential were the most important factors in deciding on a site. The agora was an open space, which acted as the city's political, economic, social and religious centre. Initially the assembly met here, but later it met in a council house (bouleuterion) or theatre, on or near the agora. Gradually the agora became rectangular with connected colonnades and stoas around three sides, and a freestanding stoa complex along the fourth side. Commerce centered on the agora and market places elsewhere, e.g. at main gates. Shrines commonly arose in the agora, as well as occasional temples. Simple houses with limited amenities and no gardens, initially arranged along haphazard and narrow streets, later became better organized in the gridiron streets system where the areas between these streets became known as insulae. Paving of streets was exceptional and sanitation poor.

From the 6th century BC city development improved progressively. Cities on the Greek mainland (e.g. Athens) remained less well planned than in the colonies of e.g. Ionia, Southern Italy, even North Africa. City states typical of central and Southern Greece did not extend to the colonies. Gradually the cities' centre of control shifted from the acropolis (which often became a sacred area) to the agora. Miletus and Smyrna in Ionia, and Syracuse and Acragas in Sicily stood out as very well planned cities. Ionian urban planning was to a limited extent influenced by the neighbouring Hittite, even Assyrian and Babylonian cultures.

During the period known as Classical Greece (5th and 4th centuries BC) orderly city planning improved further, potentiated by experts like Hippodamus of Miletus, the first city planner whose name is known to us. He reputedly planned the cities of Piraeus, Theorioi and Miletus (after its destruction by the Persians). While Greece was characterized by great cultural and scientific development, the average city dweller still experienced a very simple life, basic housing facilities, limited food and amenities, and recurrent warfare and destruction. While Athens in particular was the centre of the intellectual activity, characteristic of the 5th century BC, its town planning was quite primitive compared with cities in Ionia and the eastern colonies. There was much speculation concerning the ideal city. In outlining his concepts, Aristotle proposed that "for the sake of beauty" streets should not be planned absolutely rectangular, that defensive walls remained essential, that the agora's economic and political functions should be separated and that gymnasia should be brought from the suburbs to the city centre.

Plato had earlier advocated the gradual removal of defensive city walls (he thought the housing and street planning should be designed as defensive structures); a virtual ring of sacred temples should surround the agora, supplemented by the law courts and houses of senior officials. He also described a quite idyllic picture of safe gardens in the Ilissos valley outside the Athenian walls. The suburbs were often a bonus for the city, with many impressive public buildings, recreational facilities and farmlands, in addition to industries and cemeteries. Pausanias contrasts the agoras of previous centuries with the agoras of new cities and their sophisticated commercial facilities, shrines, temples and public buildings open to all citizens except criminals and accused persons. Streets were still rarely paved (covered with gravel, beaten earth, potsherds and seashells) but insulae (street blocks) had become square which allowed for two rows of houses, placed back to back, with direct access to the streets.

The Hellenistic era (3rd century AD onwards) built on the preceding developments but with more spectacular planning including monuments, colonnades, dramatic terraces, temples and other public structures placed in elevated settings. The gridiron street system was firmly established streets were wider and included impressive main avenues. Commercial and public buildings were now encroaching on the agora. The new cities of Alexandria and Halicarnassus were good examples of the new era, where rich patrons and kings had begun to carry the costs of the monumental structures. At this stage the Roman era dawned and with it combined Graeco-Roman developmental planning.

Before the 8th century BC, Etruria and much of central and northern Italy were inhabited by a late

Iron Age people, the Villanovans, who lived in round wattle, wood and mud huts. During the 8th age BC the Etruscans replaced the Villanovans and started building rectangular houses of stone and mud bricks. Their biggest city, Veii, was characteristic of the Etruscan culture and demonstrated quite haphazard planning with primitive or no streets, and scattered houses, temples and cemeteries. It was however; built with a virtually impregnable citadel and excellent water supply and drainage system. Through contact with Greeks in Southern Italy and the Po valley, Etruscan city planning showed gradual development, as represented in cities like Spina, Marzabotto and Bolgia. Cities like Pompeii and Capua show progressive evidence of Etruscan, Greek and ultimately Roman influence.

Early Roman planning show evidence of both Etruscan and Greek-Hellenistic influence. Military factors also affected city planning, and Roman wars of conquest soon necessitated the establishment of cities for veterans. Based on a strict gridiron system these usually rectangular cities like Augusta Praetoria and Verona were rigidly designed, largely based on the "blue print" developed in 4th century cities like Alba Fucens and Norica. With very strong central administrations they usually had excellent civic amenities, water supply and drainage systems. Much attention was also paid to religious sanctuaries. Rome, however, retained its early haphazard structure. Even after the sack of Rome by the Gauls in 390 BC, the citizens took pains to rebuild the massive defensive walls as rapidly as possible, but took no trouble to plan streets and residential areas according to newer concepts.

During the Roman Empire city planning evolved mainly from Greek Hellenistic origins, but also with a strong military component, e.g. Thamugadi and Cuicul in North Africa (1st century AD). In the provinces liked Gaul and Britain, new cities served as military strongholds but also played an important role in spreading Roman culture. Cities tended to be more compact and centralized than their Greek equivalents, to have more decoration and monumental structures, better water supplies and sewage systems, and from the time of Julius Caesar standard paving of streets. Improved building techniques and hydraulic engineering facilitated this development. Thanks to the Pax Romana, defensive city walls were no longer needed and after the 1st century AD cities required imperial permission to erect such walls. Streets of newer cities became wider, the square Hellenistic insulae tended to become oblong, main north-south streets became known as cardo and east-west streets as decumanus. The forum (replacing the Greek agora) was where main streets or avenues crossed, and housed an increasing number of temples, public buildings, colonnades and markets. It lost its air of openness. Particularly in the East the increasing size of recreational facilities like baths and amphitheatres, progressively necessitated the siting of these structures on the outskirts of cities. Cities vied with each other regarding general appearance and amenities and within cities rich citizens and nobility vied for public support (for the next election) through public donations and the dedication of buildings or other structures.

But while newer cities gained from more efficient city planning, old cities like Ostia and Rome largely retained their disorganized civic characteristics. Imperial Rome grew to a size of approximately 13 square kilometers, eventually harbouring 1, 2 million inhabitants in seriously overcrowded residential areas. The streets were always a jumble, even after urban reconstruction following the great fire of AD 64 in Nero's time. There were narrow tracts for pedestrians (itinera), wider streets for single carts (actus) or 2 carts abreast (viae), while the proper streets (vici) were close on 3m wide (allowing for projecting balconies). Julius Caesar passed strict laws to control the constant filth in the streets, but with little lasting effect. Merchandise flooded the sides of the streets until Domitian forbade the practice. Wheeled traffic was not allowed in day time, resulting in very noisy nights when wagon transport took place. At night there was also total darkness. Street lighting was extremely rare in Roman times – Ephesus and Antioch being exceptions. Between Rome's city centre (forum) and the city walls were 14 residential regions with 3 429 insulae in the New City and 2 965 insulae in the Old City, where the rich and poor lived together. Researchers estimate that there was probably on average only one domus (home of a rich person) per 26 insulae.

Villas tended to be on the outskirts of the city and in the suburbs beyond the walls. The vast majority of Roman citizens lived in multistoreyed apartment blocks (tenements) fitted into the insulae each block containing identical rooms (cenacula) with bay windows onto the street or a central court yard. The ground floor usually housed either the owner of the building, a rich tenant or tabernae (artists' shops or eating places); overcrowding and filth increased the higher upwards one moved in the building. Petronius complained about the lice. By the 3rd century BC three-storeyed apartment blocks were common, but because of increasing construction problems and the collapse of buildings, Augustus decreed a maximum height of 20m, and Trajan reduced it to 18m. But in time this was overstepped and the insula of Felicula became famous for its structures which were 5 to 6 storeys tall. In these tenements sanitation was rudimentary; the single window per cenaculum responsible for air and light had no glass panes but was covered by skin, cloth or wooden shutters. Heating was by way of individual braziers (no fire places allowed) and the fire hazard was high. Water was available at ground floor level. Insulae had their own water carriers (aquarii), porters (ostiarii) and sweepers (zetarii). Pliny commented on the common occurrence of climbing plants and potted window flowers on insulae.

Rome and other cities of the Empire became known for their elaborate recreational facilities and also sacred places. In Rome the Campus Martius, for instance, covering 200 ha, was a place reserved for the gods only, with temples, tombs, and areas for funeral pyres.

Health Facilities

The health of an urban population is of course determined by many factors which will change with time. The availability of health-related facilities such as sanitation, waste disposal, water supply and burial arrangements in Graeco-Roman antiquity is reviewed. The interest Vitruvius (early 1st century AD) showed in the potential impact of health matters (climatic conditions in particular) on the siting and planning of cities is quite striking, although his confidence in the value of haruspication (inspection of animal livers) could hardly have been productive. Four centuries earlier Hippocrates had already warned about the health hazards associated with marshy places and diseases caused by exposure to certain winds and climatic change. He advised extensively on the influence of winds, climate and seasons on urban planning.

Water Supply

An adequate water supply was essential for the siting and planning of all cities, and in early post-Mycenaean Greek cities such as Old Smyrna (11th century BC) and Pergamum, wells, springs and even rivers served as sources of water. Natural supplies could be supplemented with private and public cisterns which collected the run-off from roofs and even streets. The re-planning of Old Smyrna (7th century BC) involved the construction of a public spring house. Fountain houses and spring houses subsequently appeared in many Greek cities, both as a public amenity and an embellishment. The Greeks were not known for their hydraulic engineering, but where necessary water was indeed channeled from afar. However, the political instability of the Greek world bred a reluctance to rely on distant water sources which could be destroyed by the enemy. Piped water supplies in cities before the 5th century BC was indeed rare. Water was carried by hand. The aqueduct of Eupalinos on the island of Samos was of early Greek construction, and the Peisistratids brought water in terracotta pipes from the Ilissos stream to the agora of Athens. Before this, Athens' water supply had been limited, but Cimon also improved the supply to the agora and public places. By the 5th century BC Acragas was supplied by way of an impressive underground aqueduct. Syracuse and Olynthos had terracotta pipelines which brought water from 17 km and 18 km away, respectively. Pergamum had perhaps the most impressive aqueduct system, subsequently perfected in Roman times, which deposited water under pressure in holding tanks above the city. Water was distributed to fountains and fountain houses where it was available to citizens and some public buildings, but very few private homes. During Hellenistic times water supply to cities improved significantly. In Pergamum a law was passed which made the creation of private cisterns compulsory.

The Etruscan city of Veii was founded after the 7th century BC in a completely haphazard way in a strong defensive position, but it showed evidence of remarkably effective hydraulic engineering. Cisterns were constructed which often tapped water via rock-cut channels from underground sources. An effective rock-cut water drainage system was also present. Similar constructions pertained at other Etruscan settlements. The Romans subsequently adopted and improved the Etruscan hydraulic skills, giving it high priority in their city planning. Aqueducts became a common sight, often instrumental in revitalizing cities, and the Pax Romana decreased the hazards of enemy interference with long distance water channels. Rome's first aqueduct, the Aqua Appia was completed in 312 BC. Eventually 11 aqueducts supplied the city, the last completed in AD 226. Most of the ample water supply went to public baths, but also to public fountains, nymphaea, water basins and latrines constructed throughout the city. All apartment blocks in the insulae received water (at ground floor level), as well as some public buildings and a large percentage of rich homes (the latter carried the lowest priority). The extensive water supply system had high maintenance cost and was efficiently run as a public amenity at minimal if any profit. The aquarii were special water carriers, who distributed water from public water-points. Juvenal considered them the lowest of the slave population.

Drainage, Sewage and Waste Disposal

Evidence shows that in general the early Greeks were lax on sewage and waste disposal. There was tolerance of filth in public places, virtually no sanitation, no public latrines and few latrines in private homes. There was indeed little shame in public urination, and no waste or dung removal, which resulted in considerable open putrefaction. However, with no panGraecian co-ordination of city planning, conditions differed greatly from city to city, and there is evidence that Miletus in Ionia had a significant drainage system as early as the 8th century BC. There is also evidence that Delos had an impressive underground drainage system in pre-classical times and possibly even latrines in private houses draining into this sewer. From the 8th century BC Smyrna had cobbled streets with underground drains which emptied into the sea. By Classical times cities like Acragas, Rhodos, Pergamum and Athens had drainage and sewage systems. In Athens private owners were allowed to dig their own (covered) drains, but resultant damage to streets had to be repaired, and in due course an efficient drainage system was developed which drained streets and houses into a main channel west of the agora. Here, as in other cities, legislation dealt with illegal dumping and waste removal, but streets were often evil smelling. Inadequate or nonexistent sanitary facilities led to dumping on streets and public areas, even when statutory dumps were provided. Maintenance of streets in Athens was the responsibility of astunomoi, who had gangs of slaves at their disposal. In several cities property owners were held responsible for maintenance of streets and waste disposal. Public officials supervised this and imposed fines for non-compliance.

The Etruscans initiated effective drainage systems in order to protect their cities from regular flooding, as early as the 7th century BC. The Romans emulated them. Construction of the enormous Cloaca Maxima which drained central Rome into the Tiber river and had an exit diameter of 5 meters, was commenced in the 6th century BC. All Roman cities were supplied with drainage systems which removed excess water and human and other detritus. With few exceptions Graeco-Roman city drains simply emptied their contents outside the city walls. Homes of the rich usually had ground floor latrines which flushed into the city mains or local cess pits; in Ostia and Pompeii there is evidence of a few second floor latrines. The insulae had no latrines. Here human waste collected in commodes (sellae pertusae) was deposited into a vat under the staircase well, from where it was manually emptied into dungheaps. Urine was commonly collected in special jars and supplied to fullers who used it in their trade. However, receptacles with excrement were commonly emptied onto streets. Large numbers of water-flushed public latrines (forica) were built for the poor. A minimal entrance fee was charged. The buildings were often prominently decorated with indoor statues, and marble seats were arranged in rectangular or semicircular fashion. These forica were public in every sense – there was no embarrassment in using them, and meetings were even arranged there. Excess water from the aqueducts was utilized to flush the city sewers.

Medical Practice

In Homeric times the Greeks believed that disease was caused and controlled by the gods, with mantic and religious influences playing an overriding role. Medicine was thus in the hands of a heterogeneous community of "healers", including priests, magic healers, druggists and root cutters. During the 6th century BC a group of philosopher physicians laid the early basis of rational medicine which led to the Hippocratic contributions of the late 5th and 4th centuries BC, which abandoned religion, magic and superstition as causes of illness, thus founding empiric medicine. Simultaneously, however, the cult of Asclepius introduced elaborately designed temples for healing through incubation sleep. These Asclepieia should be seen as hydropathical institutions rather than as hospitals in the modern sense of the word. They were however seldom placed in cities - like modern health spa's they were usually constructed in a scenic country site. The typical Hippocratic doctor, on the contrary, probably practiced from very basic facilities in the city – perhaps a consulting room (iatreion) which was also his living quarters. A limited number of salaried civic doctors existed from at least the 5th century BC but we have no information about their facilities. Hospitals as we know them did not exist. Basic operations, e.g. "cutting for stone" (lithotomy) were performed in the consulting room or the patient's home. The so-called "medical schools" of Cos and Cnidos (5th century BC) were not circumscribed building complexes as we know them today; in fact, tradition has it that Hippocrates taught under a plane tree. The "medical school" of Alexandria (late 4th century BC) functioned in the city, but the common concept that it was associated with the famous Museion and/or library cannot be substantiated. Medical education at the time consisted of in-service training and was not really institution bound.

Roman general medicine represented a continuation of the Greek model, but was significantly influenced by military implications. The first true hospitals were military institutions (valetudinaria) erected at the war front. There is evidence that in time valetudinaria were also instituted for imperial staff and the upper class, and that there were "hospitals" for gladiators and slaves on large farms (latifundia). Doctors still practiced from small consulting rooms (tabernae), some of which were sited in public baths. The suggestion that some tabernae developed into small house-hospitals (nosocomia) is little more than speculation; well-to-do households did however have quarters for the sick. The early Christians soon became known for their selfless dedication to the relief of suffering. When Constantine (4th century) granted religious freedom and fostered the Christian belief, shelters for paupers, pilgrims and church messengers in the Greek-speaking East gradually evolved into the earliest hospices, called xenodocheia. In an attempt to gain support for his pagan countermeasures, Julian the Apostate promoted the creation of xenodocheia which then also spread to the western part of the Roman Empire. Christian monasticism resulted in the creation of so-called infirmaries attached to monasteries. Mainly the result of Christian benefaction during the 4th and 5th centuries, hospitals now appeared in many centres.

Burial and Cemeteries

A belief in pollution by the dead (cultural and religious rather than physical pollution) held sway throughout Greek antiquity. Contact with the dead was thus minimized and within the city walls there were no structures for the handling of the dead. Before the 8th century BC citizens could still be buried inside city walls but not after this time. Solon's extensive legislation (early 6th century BC) regarding burial covered the prothesis (preparation of the body in the house where he/she died), the ekphora (funeral procession, before daybreak), burial and/or cremation procedures, and rules of mourning as well as follow-up visits to the grave. This remained in effect up to Roman times. Originally graves were simple square holes, with a minimal grave mound, but during the 5th century elaborate grave stones, stelai and other structures started appearing on graves, until severely curtailed by decrees of Demetrius of Phaleron in the 4th century BC. Slaves, strangers and the indigent were afforded decent burials, but the bodies of criminals were tossed into the sea, stone quarries or special pits. Persons who committed suicide had their right hands chopped off before being buried. Cemeteries commonly arose alongside streets leaving the city gates.

Roman concepts of death pollution showed similarity with those of the Greeks, but hygienic considerations gradually played a bigger role than the religious cultural components. The laws of the Twelve Tables laid down legal requirements for handling of the dead and corresponded closely with Solon's laws in Athens. Because of regular epidemics and the progressive growth of Rome as a city (approximately 750 000 inhabitants in the 1st century BC and more than a million in the 2nd century AD) proper management of the dead became a major issue. After the 5th century BC burials and cremation within the city walls were allowed only in exceptional cases. Outside the Esquiline gate in the grove of Libitina a funeral industry developed, run by professional libitinarii and funerarii who managed the vital burial process. In due course two burial areas developed: the loca religiosa along the highways leading from the gates of Rome, where proper graves, often richly decorated and family concerns with a distinctly religious character, were located, and the local public outside the Esquiline gate which served as place of execution and the final resting place for indigents, criminals and unidentified bodies. Here cadavers were commonly left in open mass graves (puticuli) which were covered over from time to time. Augustus finally converted this area into gardens and it is uncertain how the bodies of criminals and the poor were subsequently handled. Mass cremation in public crematoria (ustrinae) could have partially solved the problem, but evidence is that burial gradually replaced cremation in popularity after the 7th century. On the Campus Martius, also outside the city walls (pomerium), a number of emperors and other illustrious individuals were buried in a sacred area with many temples, shrines and monuments.

Sport and Gymnasia

The Olympic games introduced in 776 BC typified the Greek approach to sport, health and fitness. Early sports grounds were called gymnasia, usually constructed outside the city walls as a public facility for all citizens. Commonly placed in sacred groves beside streams, there were three such gymnasia outside Athens. They typically contained running tracks, jumping pits and areas for throwing the javelin and discus. Two of the Athenian gymnasia were big enough for riding lessons and cavalry exercises. Other amenities included dressing and washing rooms, dusting rooms for powdering before exercises, rooms for ball games and boxing and palaestrae (low buildings mainly used for wrestling training). They were commonly used by epheboi (postpubertal boys who as part of their military training were also given cultural education with lessons in philosophy and literature). In Athens gymnasia employed trainers (gymnastai, paidotribai). Sporting events were also held inside city walls and in the agoras in particular. At Pergamum there were interconnected gymnasia situated on the high terraces as part of a complex of temples, shrines and fountain houses. In late Hellenistic times these gymnasia were gradually replaced by the Roman public baths.

The Greek concept of sport and athleticism was never really accepted by the Romans. The younger Pliny clearly disapproved of it. Sulla, Pompey and Julius Caesar attempted unsuccessfully to popularize Greek games. Subsequently Augustus introduced his four-yearly Actiaca games, but as was later the case with Nero's Neronia, it never caught on. In AD 86 Domitian built the Circus Agonalis (15 000 seats) for his Greek-styled four-yearly Agon Capitolinus events, and the Odeon for spiritual and cultural contests. These were reasonably successful for selected audiences and lasted into the 4th century. For the Roman upper classes Greek games were associated with nudity, even immorality and never rivaled their enthusiasm for gladiatorial shows and horse races. However, Greek communities persisted with their own games, which flourished in Attica and parts of southern Italy – but never in Rome.

Public Baths

Public baths, referred to as thermae, was a Roman institution introduced during the 2nd century BC. Dependent on generous water supply, these baths were as much a social recreational facility as an hygienic one. Initially males and females were admitted separately and a minimal admission fee (1/4 as) was required by the owner-builders; children were allowed in free of charge. By 33 BC Rome had 170 such public baths and they were steadily increasing in number. That year Agrippa, the aedile who was in charge of Roman baths, generously proceeded to pay all bath fees, and also built a new bath which was free of charge. His bath set the new norm and also became an architectural model. In the 1st century AD a succession of emperors (Nero, Titus and Trajan) built baths, ever growing in size. The baths of Caracalla, 27 acres, was eclipsed by Diocletian's complex covering 32 acres; Constantine built the last such bath in the 4th century. Ultimately Rome had in excess of 800 public baths.

The baths, built by all significant communities throughout the Empire, had a common basic lay-out consisting of changing rooms, a sweating room (caldarium) heated by an under-floor hypocaust, and a cold (unheated) room (frigidarium), both with plunging pools. More elaborate baths had dry heat rooms as well (laconica), intermediate temperature rooms (tepidaria), swimming pools, massage rooms, latrines, gardens and colonnades, service rooms and stoke-holes (praefurnia), lecture halls, libraries and museums, as well as exercising areas which largely corresponded to the Greek gymnasia. The baths played a central role in the daily lives of a large proportion of Romans. In Domitian's time men and women were bathing together, but in time this caused so much scandal that Hadrian again separated the sexes in AD 138. Public baths were constructed inside the city walls, as well as in the suburbs.

Recreation and Entertainment

Theatres

The theatres in which Greek drama flourished in the 5th and 4th centuries BC were unroofed structures, usually situated in a hollow or against a hillside in order to accommodate the semicircular tiers of stone seats for the spectators. This component was called the auditorium (theatron), and it looked down on the circular or oval orchestra with its beaten ground floor for the performers. Primitive early examples are to be found at the Attic towns of Icaria and Thoricus, while Epidaurus had a splendid, near perfect theatre build in the 4th century BC. Athens's original theatre was next to the agora, and the subsequent theatre of Dionysus against the Acropolis hill was built in the late 6th century BC. Most Greek cities had their theatres, with the oldest probably in Thoricus (south Attica), 10th century BC.

In the Roman world the oldest theatre was probably that in Pompeii, 5th century BC. Theatres gradually ceased to be open-air structures and became roofed and enclosed in buildings. The seating was no longer against a hillside but incorporated into brick or stone walls. Examples of these were the Theatre of Pompey in Rome with a semicircular diameter of 160 meters and seating for 27 000 spectators; the Theatre of Balbus, with 7 700 seats, and the Theatre of Marcellus with 14 000 seats and a diameter of 150 meters. During the Republican era Greek-styled drama was still as popular as the circus races, but then it went into decline. The last significant Roman dramatists like Accius, Pacuvius, Plautus and Terence wrote in the 2nd and early 1st century BC. Although drama was still performed well into the Christian era, diverbia (dialogues), cantica (songs) and pantomimes became more popular. The brutal culture of the amphitheatre and its munera (games) also found its way into the plays which became lurid, even bloodthirsty. This eventually led to its demise. The Theatre of Marcellus closed in AD 235, but some theatre activity continued up to the 6th century.

Amphitheatres and Circuses

The satirist Juvenal remarked that the degenerate Roman citizens cared only for bread and circuses. The emperors arranged for the regular distribution of free bread, and also paid for frequent public amusement in the theatres and amphitheatres. Public holidays featured frequently on the

Roman calendars. At the end of the first century BC there were 59 holidays per year. At the time of Julius Caesar this had increased to 132. By AD 354 there were approximately 200 holidays per year, 175 of which were dedicated to games and races. Providing adequate facilities for these events was a major challenge for city planners, and in the heyday of the Empire the Roman citizens enjoyed more spectacles than any other community has, before or since.

The Circus, designed primarily for the races, was a long rectangular structure, rounded at one end into a semicircle. The Circus Maximus in Rome, oldest and largest, was built in the 2nd century BC but enlarged by Caesar and Trajan to an ultimate size of 600 x 250 meters, seating 225 000 (some said 385 000) spectators. The Circus Flaminius was 400 x 260 meters, and the Circus Gai (1st century AD) 180 x 90 meters. The smaller circus of Maxentius was outside the city walls on the Via Appia. Down the centre of each arena was the spina with turning posts at each end, dividing the area into two runs. Circuses were also found elsewhere in Italy and in the provinces. In Rome the initial 12 races per day (7 laps each) were later increased to 24 races per day. There were horse races and chariot races (up to 10 horses per chariot) and they competed under team colors – initially red and white, and later blue and green as well. The events were tremendously popular with the masses and were associated with heavy betting. However, persons like the younger Pliny and Marcus Aurelius expressed their disdain with the races.

Gladiatorial shows and spectacles with wild animals were held in amphitheatres, which were round or oval structures. These very popular performances, characterized by brutality and bloodshed, originated in Rome during the 2nd century BC, and were initially held in temporary wooden amphitheatres, even in circuses. The first permanent amphitheatre in Rome was constructed in 29 BC, burnt down in AD 64, and was then splendidly reconstructed as the Flavian Amphitheatre or Colosseum. It was oval, 188 x 156 meters in diameter and the seating area four storeys tall, accommodating 45 000 spectators. Underground were extensive passages and rooms, and the central arena (3 500 meters) could be flooded for naval displays (naumachia). There was also the smaller amphitheatre Taurus, constructed by Augustus for land combats, and two Naumachia's (557 x 536 meters) specifically designed for naval combats. Amphitheatres were also built elsewhere in the Empire. The shows, called munera, were mainly financed by the emperors, and admission was free. The first permanent amphitheatre of Roman construction was probably that at Pompeii, 80 BC.

The gladiators were recruited mainly from prisoners of war, convicts and men condemned to death by the courts. They were trained in special schools (ludi). Munera, like the races, were dawn to dusk events, in which pairs of gladiators would fight each other to the death. Shows often started with mimic fights (prolusio), preceding the real battle (hoplomachia); group combat and naval battles were also included. The munera involved wholesale massacres. Records from the 2nd century mention two relatively minor munera where 202 and 350 gladiator pairs respectively took part, and a major munus, lasting 117 days in which 4 741 gladiator pairs were involved. Trajan, Hadrian and Marcus Aurelius scaled down the bloodshed, and in the 5th century the arrival of Christianity put an end to gladiatorial combats in the West. Wild animals often formed part of the shows, but gladiators never fought animals. This was left to bestiarii. Criminals who were convicted "ad bestias", were fed to the animals – and many early Christians died this way. When Titus inaugurated the Colosseum in AD 80, 5 000 animals were killed on one day, and we have records of two munera in Trajan's time where 2 256 and 443 animals were slaughtered. In 326 Constantine abolished "ad bestias" convictions, and ordained that large numbers of criminals be sent to work in mines, thus abolishing a main source of gladiators.

Public Gardens and other Facilities

Mumford states that early Greek cities up to the 5th century BC had very few gardens and parks. But by the 5th century BC Athens had regular gardens planned and trees planted, and Plato presents an idyllic picture of the Athenian suburbs on the banks of the Ilissos river. Suburbs had also acquired public amenities like libraries, bath houses, even museums. The new city of Alexandria represented the Hellenistic era with carefully laid out gardens, public buildings, a museum and library. During the Roman Empire water supplies to cities received high priority, and in its heyday Rome had approximately 500 fountains, 30 public parks, 8 grassy commons and many gardens. City planning in the Roman provinces usually included public facilities like libraries, gardens or parks.

Mesopotamia and Nile Valley

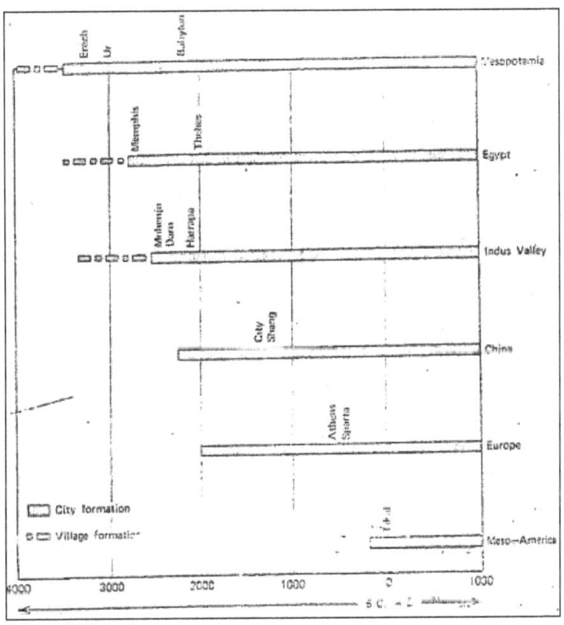

There emerged two points of view on the origin of city; one there was a single area of the origin of cities – a single culture hearth. The cities have originated in the fertile crescent of Mesopotamia. The city formation was diffused from Mesopotamia to the Nile valley, The Indus valley, the Hwang Ho valley and perhaps to Meso-America. The traits that have been collected so far, such as deities, an exalted ruler & central authority, great temples, division of caste and labour, the beginning of writing and calendar matches. But those who disagree argue that these traits are too specific to have been spontaneously developed. So, they must have originated in a common culture hearth and have been modified to suit each area to which they were diffused. Another approach towards origin of cities states that cities originated "spontaneously" in different regions without external influence, even though the indications are that cities in different regions were formed as much as several millennia apart. The lines of communications between the various areas involved were supposedly non- existent and transportation was not developed at the scale needed for migrationor travel between those areas. The environment within which cities developed in Mesopotamia, Egypt, the Indus Valley and the Hwang Ho valley were mostly similar, in being the level floodplains and soils & climate favouring surplus agricultural production. But the cities of Meso-America were first developed in what might be considered as harsh environment. The agriculture was challenging, transportation was crude and many elements of the technology of the Old World were absent.

A system of villages developed first in Mesopotamia. The main concentration of people, around 200-500 individuals, was in the hills and along piedmont of Tigris & Euphrates river and not along flood plain proper. These included the earliest cities of Mesopotamia like Erech, Eridu, Ur, Lagash and Larsa in the southernmost part of the Tigris-Euphrates river valley, and Kish & Jemdet Nasr in the midsection of the same river valley. All of them had transformed from villages to cities. But the process of transformation was gradual and took centuries, before it was complete. Each city-state was ruled by the king, who was also its chief priest. It was mandatory for the farming population to return the king the part of their surplus agricultural output. The surplus was stored in the main temple. It formed the main source of sustenance for the ruling group of the city. The cities founded at this time were on sites located on the interfluves between channels of the Tigris and Euphrates River system where they were afforded the greatest protection from floods that occurred frequently. The earliest cities grew up within the walled section that had temples, and formed the characteristic feature of Mesopotamian cities. The main dominance was of the priestly class on the social structure. They guided both economic and ritualistic aspects of the society. The temple palace complex was surrounded by a wall.

It had a dual role, one to act as a defence against marauders from adjacent city-states and two, provided protection against the recurring floods from Tigris- Euphrates valley. The walls were made of fired and unfired mud bricks for most of the houses, temples and palaces. The arrangements of the houses were irregular. They were interrupted by open spaces in the vicinity of temples and other public buildings. Streets were narrow lanes without surfacing or proper drainage. The refuse accumulated on the lanes. The excavation of Ur reveals one of those extreme conditions where the refuse had gradually raised the level of the streets to such an extent that houses had to create entrance on the second level, since the original floor was below the street level. The chapels, small shops, and rows of booth were interspersed among the cluster of houses and were supposedly used by merchants and artisans. The outer margin of the city was a zone where the poorest elements of the population lived in mud and huts. Farming community were settled in the periphery of the city. The cultivation of the crops within or near the margin of these cities was beneficial in that it facilitated the movement of produce to markets in the heart of the city as well as providing a more assured food supply.

The early development was also marked in the valley of the lower Nile River. The agglomerated settlement of the Nile valley dates from 3500 to 400 B.C. These early settlements were crude walled clustering with the dwellings like huge barracks and narrow lanes between that served as sewers as well. As in the Mesopotamia, in the Nile valley too, the development of the walled villages took place in the delta area of the lower river valley. The villages were clustered into politically independent units. Each had its own cooperative irrigation systems for cultivation of staple grains like wheat and barley. Animals were domesticated and metallurgical tools were used. In the social sphere, there was stratification and an emerging division of labour. The earliest cities that emerged in this region about 3100 B.C. were Memphis, This, Heliopolis, Nekhab (El Kab) and later Thebes. The remnants of these cities have mostly been covered by silt deposited in the annual floods of the Nile Valley. The thought on diffusion further gets credence from the certain artefacts found on the sites of the ancient Nile River, as they are believed to have been borrowed from Mesopotamia. It also includes the concept of early writing. There had been a more gradual transition from village to city in the Nile valley than in Mesopotamia.

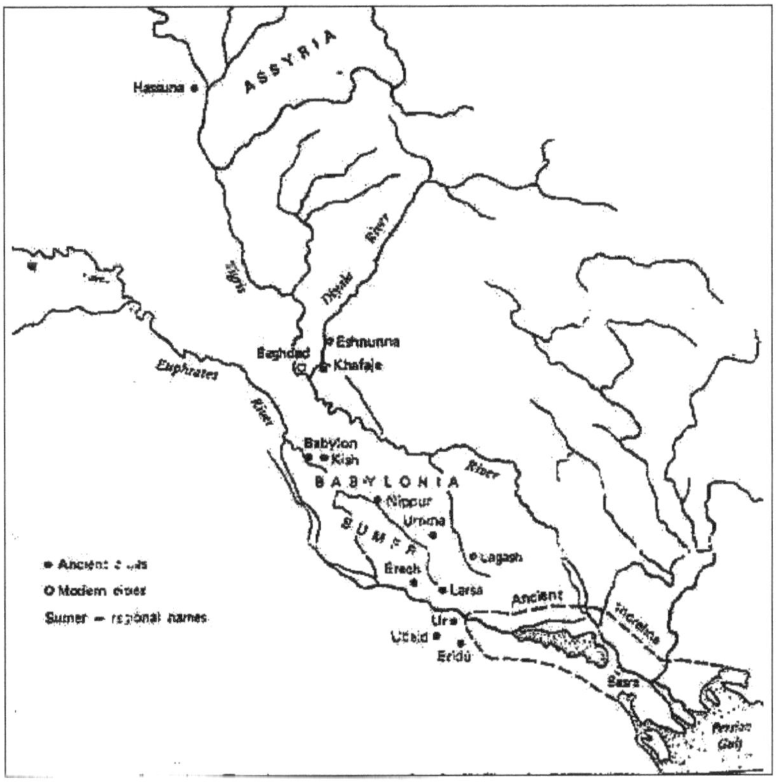

Ancient Mesopotamia.

The practice of changing the site of the capital may have resulted in the conspicuous absence, in development of large cities. Around 2900 BC Akhetaton became the first capital while Memphis enjoyed this status after 2700BC. After about 2000BC Thebes became a large city and Akhetaton served as a capital for a short period. Part of the reason why traces of earlier cities in the Nile Valley are lacking may have to do with the building materials used. The type of building material used may be one of the reasons for the missing traces of the earlier cities in the Nile valley. The materials do not readily withstand the weathering of time were used like sun dried clay bricks supplemented with wattle and daub. Later too, when stone was used for construction of the pyramids, the houses of the people were made of perishable materials. The remnants of cities that developed after 2000 BC indicate that multi-storeyed houses were fairly common. Social stratification with the slum area, a working men's section (handicrafts), and a core area including a palace, temples, store-houses for the surplus food and public buildings. The streets were un-surfaced with no provision of drainage, like those of the Mesopotamian streets.

A third area that marked the population agglomeration originated was the upper Indus River Valley. There are lots of evidence to support existence of numerous towns and villages in this region. The region was based on a highly productive system of agriculture that facilitated in emergence of these towns and villages. The system of irrigation and water impoundment was most likely a sophisticated system. The beginning of the settlements in this area dates from about 3300BC although cities as such dates from about 1500 -2500 BC. The cities of Mohenjo-Daro and Harappa, in the Indus Valley, developed into major cities by 2500 BC. The city of Harappa was in Punjab at the headwaters of the Indus. Later, these cities rose to the status of capitals of an empire that flourished between 1500-2500 BC. Agriculture was supported by system of irrigation and surpluses of wheat and barley were produced in this area. The people of the region knew use of tools as well as domesticated animals and used wheeled vehicles.

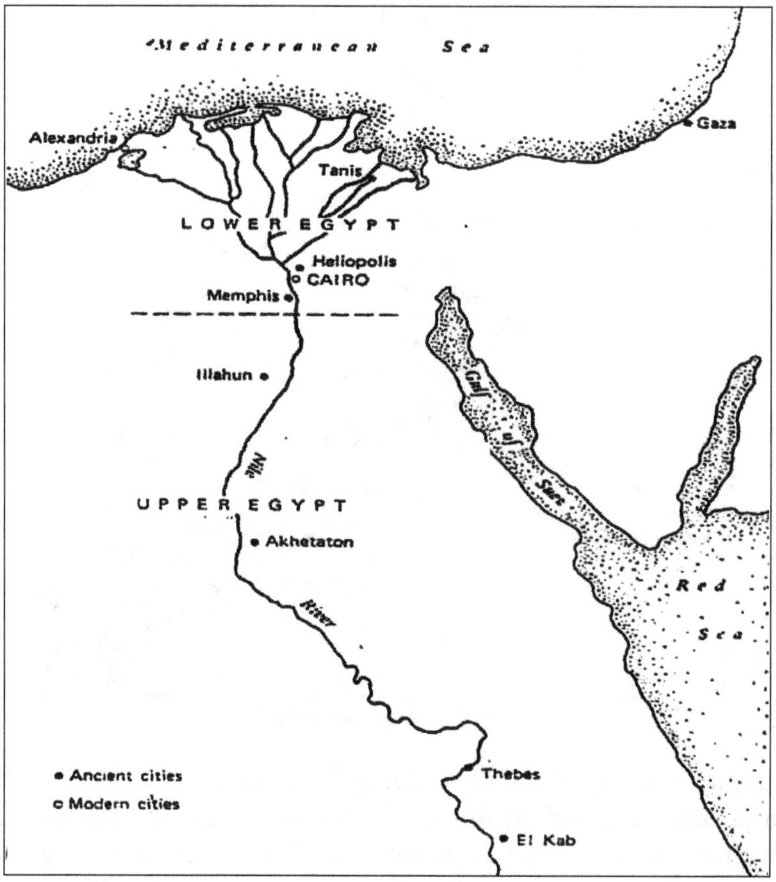

Ancient Egypt.

Another area that witnessed the origin of earlier cities was the Wei Valley, not far from the Hwang-Ho River in China. Towns were established about 2250 BC in this region. The level of technology was lower than in other areas of city origin. There was no evidence of use or knowledge of metallurgy. The people produced surpluses of food like millet, and domesticated animals. The first cities to have developed in China seem to have been in Hwang Ho Valley, with the earliest being the city Shang (Yin) on the Hwang Ho at Anyang. Shang civilization was based on agriculture that included cultivation of wheat, barley, millet and rice. There seems to have been water impoundment measures & canalisation that resulted in increasing the productivity staple food-stuffs. There are evidence of domestication of animals and wheeled vehicles. The power structure was also well defined with a hereditary political leader and nobility.

The city of Meso-America appears to have originated centuries preceding Christian era. The advanced communities had been developed in the pre-Columbian times by the peoples indigenous to this region notably the Maya, Zapotees, Mixtecs and Aztecs. The early cities of Meso-America such as Tickal and Uaxactun, and the later like Copan, Mayapan supposedly were developed by a society that produced surplus of maize. There was development of class structure. The leadership was providedby the priestly class.

None of the cities that developed in this region exists today. Some were said to be submerged beneath flood waters, some gradually disintegrated and others were buried by the accumulation of sand and silt. The earliest cities did not occupy large amount of land area nor did they have populations that today would be considered large. Kaekemish on the Euphrates River covered 240 acres and Mohenjadaro on the Indus River covered about 600 acres, Ur with its canals, harbours and temples occupied 200 acres, with its population as high as 24000 people. Nineveh included about 1800 acres of land area. Crowdedness, congestion, slums and pollution had been the urban problems since these cities were first formed.

Medicine in Ancient Mesopotamia

Mesopotamians somehow knew the natural origin of numerous illnesses, which may occur of over-exposure to heat or cold, overeating, eating spoiled foods, or drinking too much of an alcoholic beverages. The fact that some diseases were contagious was certainly recognized, but there was obviously no perception of how contagion occurred. In general, however illness was perceived as an intrusion coming from outside a person and entering the body in some fashion. It may be that this concept helps to explain the use of purgatives and enemas even for illnesses that do not involve the digestive system or for complaints that we would tend to see as psychological in original.

The illnesses were often blamed on pre-existing spirits including gods, ghosts, etc. Each spirit was held responsible for only one illness in one part the body. So, usually the hand of god of stomach corresponded the diseases of stomach. Some diseases simply were identified by names "bannu" for instance. Also, it was recognized that various organs could simply malfunction, causing illness. Gods could also be blamed at a higher level for causing named diseases or malfunctioning of organs, although in some cases this was a way of saying that symptom X was not independent as usual, but was caused in this case by disease Y. It can also be shown that the plants used in treatment were generally used to treat the symptoms of the disease, and were not the sorts of things generally given for magical purposes to such a spirit. Presumably specific offerings were made to a particular god or ghost when it was considered to be a causative factor, but these offerings were not indicated in the medical texts, and must have been found in other texts.

Mesopotamian mother and child, probably a fertility image.

Remedy

In Mesopotamia, healing remedies have taken an enormous variety of forms; from preparation using plants, animals, and minerals, to remove of infected or diseased parts, to prayers or incantations for spirits or gods to restore health. The therapeutic medical texts combined the two types of treatment, the magical and the medical cures. Prayers, recitation, sacrifices were common religious means of beseeching the gods for a cure; however, along with these a real pharmacopoeia of drugs was used in the treatment of disease. In addition to clay tablets which report illnesses with their symptoms and diagnosis, prognosis, and treatment, others were found that list drugs and their appropriate uses. Many plants, minerals, and animal substances were the therapeutic agents, and were given by mouth in compositions, applied as salves and fomentations, blown into orifices, inhaled as vapors and fumigations, and inserted as suppositories and enemas. Oil was apparently the principal balm for open wounds, probably preventing the adherence of overlying dressings. The medications were administered according to rituals, the time of the day, and the positions of constellations.

Very many elaborations could be given, for example, quantities of specific ingredients, the time of day, and the number of days to administer the medication. In a few instances, warnings on toxicity of certain ingredients are given, An example of a prescription is the followings: "If a man's tongue is swollen so that if fills his mouth, you dry tamarisk leaves, leaves of the adāru- plant, leaves of "fox-grape," (and) "dog's tongue"-plant; you chop them up finely and sift; you knead them with juice of the kasū- plant; you rub the top of his tongue with butter; you put (the medication) on his tongue, and he will get well."

Particularly in the case of magical treatments, instructions may include such specifics as "such-and- such plant to be pulled up before sunrise," or "set outside under the stars overnight." Containers normally made of leather were often included in instructions for magical treatments. They might contain, for example, hair from a black dog, a piece of a dirty (menstrual) rag, or pig manure to be worn around the neck. The corpus of medical texts, being in large measure a compilation of material from what may have been separate corpora of texts of the magical and medical treatments, contains treatments from both sources mixed together. A considerable number of the medical texts from Assur have colophons showing that they belonged to a man named Kişir- Assur, who identifies himself as an exorcist of the temple of the god Assur. If he was not merely a bibliophile but made professional use of these tablets, it would appear that there was considerable fluidity and overlap in the use of prescriptions by both categories of healing professionals.

Practitioners

Medical practice probably was in the hands of three types of priests; baru, ashipu, and asu. The baru as a diviner dealt with diagnosis and prognosis, but not only of illness. He also had to discover the causes and probable outcome of many other kinds of catastrophe. The ashipu, as an exorcist who drove out evil demons, was called on to rid a house, a farm, an area, and also sick people of occupying spirits. The asu (the name of Biblical king Asa or Asa-El, "healer of god," may have derived from the Babylonian asu) apparently acted as a physician, employing charms and divination but also drugs and operations. The healing priests received their education in schools that were associated with the temples. The source of their learning, in addition to practical instruction, was the large number of tests available in the form of clay tablets. By the seventh century BC, for instance, the library of Ashurbanipal contained over twenty thousand tablets, which were only discovered about a hundred and fifty years ago at the site of ancient Nineveh. They are still the most extensive source of knowledge about Mesopotamian society, including medicine, but recently tablets have been unearthed that date back to Sumerian times.

The priest- physician ministered mainly to the court, nobility, and upper classes, but apparently there were also barbers who performed some surgical procedures and did the branding of slaves. Veterinary practice may have been handled by either the low- class "doctors of oxen or asses," is not known.

Fees and Punishments

The Code of Hammurabi devotes ten statements out of 282 provisions to the fees due medical practitioners and their punishments for failure.

- If a doctor has treated a freeman with a metal knife for a severe wound, and has cured the freeman, or has opened a freeman's tumor with a metal knife, and cured a freeman's eye, then he shall receive ten shekels of silver.
- If the son of a plebeian, he shall receive five shekels of silver.
- If a man's slave, the owner of the slave shall give two shekels of silver to the doctor.
- If a doctor has treated a man with a metal knife for a severe wound, and has caused the man to die, or has opened a man's tumor with a metal knife and destroyed the man's eye, his hands shall be cut off.
- If a doctor has treated the slave of a plebeian with a metal knife for severe wound and caused him to die, he shall render slave for slave.
- If he has opened his tumor with a metal knife and destroyed his eye, he shall pay half his price in silver.
- If a doctor has healed a freeman's broken bone or has restored diseased flesh, the patient shall give the doctor five shekels of silver.
- If he be the son of a plebeian, he shall give three shekels of silver.
- If a man's slave, the owner of the slave shall give two shekels of silver to the doctor.
- If a doctor of oxen or asses has treated either ox or ass for a severe wound, and cured it, the owner of the ox or ass shall give to the doctor one sixth of a shekel of silver as his fee.

A notable feature of the Code is that there is here enshrined a code of ethics, which raised the standard of medical practitioner to a very high level, first by fixing the remunerations which they are to receive at a substantial figure, and secondly by punishing them for transgressions due to ignorance. As such transgressions were surgical errors, a by- product of this code was the virtual suppression of surgery. To this high standard the physicians of ancient Iran succeeded and from this high standard they never fell.

Code of Hammurabi

Code of Hammurabi, collection of laws promulgated in the early second millennium BC by Hammurabi, King of Babylonia son and successor of Sin- muballit of the Amorite dynasty. He owed his success not only to military ability but also to attention to administration and economic justice, reflected in his correspondence and his laws. Most of the Code of Hammurabi is extant on a diorite monument, 8.2 ft. high, discovered at Susa (now Shush, Iran) in 1902 and now in the Louvre, Paris. At the top of the stele a relief shows Hammurabi paying homage to the sun- god Shamash, also the god of justice; below is engraved the cuneiform text, in finely executed writing reading vertically from top to bottom in a series of horizontal bands. Portions of the text are extant on clay tablets from Babylonia and Assyria from periods down to the middle of the 1st millennium BC.

The text comprises (1) a prologue telling of Hammurabi's divine call and his piety, (2) the corpus of laws, and (3) an epilogue extolling Hammurabi's establishment of justice in the land. The laws, arranged largely according to subject matter, deal with such topics as offenses against the administration of justice and property, land tenure, trade and commerce, marriage, family, and inheritance, assault, regulations governing surgeons and builders, agriculture wages and slavery. Penalties included death by drowning and impalement, mutilation, scourging, banishment, enslavement, and the payment of reparations.

Similarities between certain laws in the Pentateuch (first five books of Bible: Genesis, Exodus, Leviticus, Numbers, and Deuteronomy, which contain the major collections of ancient Israelite laws and are frequently called the Law, or Torah) and those of Hammurabi formerly led to the suggestion of direct borrowing by Moses. This theory has been discountenanced by the discovery of other collections of Mesopotamian laws, which show that Hammurabi was neither the only nor the first lawgiver in the ancient Near East.

Though commonly referred to as a code, Hammurabi's laws have been shown to be not a codification of statute law, but a collection of specimen decisions given by Hammurabi as judge, arranged in the literary form, "if so- and- so happened, so- and- so will follow," known as "omen- form." The epilogue recommends the judgments and decisions recorded on the stele to future rulers and people.

The contents of the "code" show that its authors were familiar not only with the specific decisions incorporated, but also with earlier related documents such as the "code" of UrNammu and Lipit-Ishtar.

The Code of Hammurabi is engraved in 49 columns and sandwiched between a prologue and an epilogue in a rhyming style giving the titles and achievements of Hammurabi and proclaiming the authority of his laws. The middle section is divided into 282 codes of Law by the modern scholars. The subject matter of the Code are:

Administration of Justice

1-5- False witnesses; corrupt judges.

Offences Against Property

- 6-14- Theft, including kidnapping of a minor.

- 15-20- Runaway and stolen slaves.
- 21-25- Housebreaking, rubbery, looting.

Land and Houses

- 26-41- Tenure of land held as a fief from the king.
- 42-48- Cultivation of arable land by tenants.
- 49-52- Regulations governing the financing of tenant- farmers.
- 53-56- Offences arising from negligent irrigation.
- 57-58- Trespass of cattle on corn- land.
- 59- Unauthorized cutting of trees.
- 60-65 & A- Cultivation of palm- plantations.
- B-E, G, H- Regulations concerning renting, building and repairing of houses.
- F, J, L- [Fragmentary].

Polished black diorite stele found near Susa, Iran, with Code of Hammurabi inscribed and bas- relief at the top showing the king of Babylonia receiving laws from sun god Shamash. The monument on which is engraved most of Hammurabi's Code discovered in 1902.

Merchants and Agents

- L-R- Loans from merchants' interest rates.
- S,T- [Fragmentary].
- U- Division of profit and loss between partners.
- 100-107- Regulations governing transactions of merchant's agents.
- 108-111- Regulations concerning (female) inn- keepers, (who apparently also served as small brokers).

- 112- Fraud by a carrier.
- 113-117- Distraint for debt.
- 118-119- Delivery of dependents into bondage for debt.
- 120-126- Deposit of goods.

Women, Marriage, Family Property and Inheritance

- 127- Slander of high- priestess or married woman.
- 128- Wifehood; cohabitation without contract does not give a women the status of wife. 129-132- Adultery.
- 133-136- Remarriage of wife: this permitted in the husband's absence only in certain circumstances, though unconditionary for voluntary desertion.
- 137-143- Divorce.
- 144-149- Concubinage and a circumstance in which bigamy is permitted.
- 150- Inheritance of property settled by husband on wife.
- 151-152- Liability of spouses for debt.
- 153- Penalty for wife who murders husband impalement.
- 154-158- Incest.
- 159-161- Breach of contract after betrothal.
- 162-164- Disposal of dowry after death of wife.
- 165-167- Division on inheritance amongst sons.
- 168-169- Dishersion.
- 170- Legitimation of sons of concubine.
- 171-174- Rights of concubine and non- legitimated sons; widow's property.
- 175-176- Rights of free woman married to a slave.
- 177- Disposal of previous husband's property at re-marriage of widow.
- 178-184- Rights of cult women to dowry or paternal inheritance.
- 185-193- Adoption.
- 194- Substitution of changeling wet- nurse.

Assault and Personal Injury

- 195- Assault by a son on his father.

- 196-205- Penalties for inflicting bodily harm- talon when the victim is freeman, a fine when is a villain or slave.
- 206-208- A man inflicting accidental injury is subject only to a fine or costs, not to talon.
- 209-214- Assault upon a woman resulting in miscarriage.

Professional Fees and Responsibilities

- 215-225- Fees payable to surgeons and veterinary surgeons for successful operations, and penalties incurred where result of an operation is unfavorable.
- 226-227- Penalties for existing of procuring the excision of a slave mark.
- 228-239- Fees for builders and shipmasters penalties for negligence.
- 240- Collision between a galley and a sailing ship.

Agriculture

- 241-249- Distraint, hire death and injury of oxen.
- 250-252- Responsibility for damage by goring ox.
- 253-256- Wrongful conversion by a bailiff.
- 257-258- Hire of ploughman an ox- herd.
- 259-260- Theft of agricultural implements.
- 261- Hire of stockman or shepherd.

Rates of Hire

268-277- Rates of hire of animal, wagons, laborers, craftsmen, ships.

Ownership and Sale of Slaves

- 278-279- Liability for slave sold when suffering from sickness or subject to a legal claim.
- 280-282- Re- claiming of slaves bought abroad and returned to Babylonia.

Three Articles in Mesopotamian Pronunciation and English Language

- Šum- ma a w- wi- lum i- in duma a- wi- lim úh- tap- pí- id i- in- šu ú-ha- ap- pa-du šumma awīlim īn mār awīlim uhtappid īnšu uhappadū.
    - Article 196: If someone causes loss of the eye of an aristocrat, then he deserves to have his eyes destroyed.
- Šum – ma gir-pad- du a- wi- lim iš- te- bi- ir gir- pad- du- šu i-še- eb- bi- ru šumma esmet awīlim ištebir esmetšu isebbiru.

- Article 197: If he breaks someone's bones, then his bones should be broken.

- Šum- ma i- in maš- en- kak úh- tap- pí- id ú lu gir- pad- du maš- en- kak iš- te- bi- ir 1 ma- na kù- babbar i- ša- gal šumma in muškenim uhtappid u lu esmet muškenim ištebir 1 mana kaspam išaqqal.
  - Article 198: If he destroys the eyes of a layman or breaks the bones of a layman then he should pay one silver enamel.

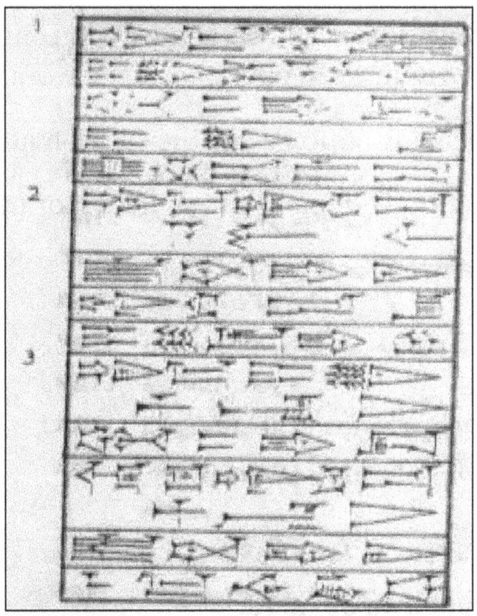

A portion of the Code of Hammurabi in cuneiform.

Medical Documents and Drug

In addition to magic and sorcery, Mesopotamians used herbal, mineral, animal products, and drugs for medical treatment. To demonstrate fever and febrile status, they drew a chafingdish. The prescriptions on Sumerian steles reveal that Lulu, a physician who practiced medicine in Ur, ancient Sumerian and Babylonian city in southern Mesopotamia, practiced medicine without the help of magic and sorcery. A Sumerian physician who lived over 3000 years ago, has left a twelve-page medical document inscribed on a stele, "15.9 × 9.6" centimeters. The medical documents were interpreted in 1953. They reveal that Sumerian physicians classified drugs into three categories including herbal, mineral and animal products. The herbal drugs included cassia, asafetida, myrtle, time, willow, pear, and spruce. For making herbal drugs, they used seeds, roots, branches, gum, resin, and fruits of various plants and trees. Different powders, ointments, drops, and suspension were also used. Mesopotamians mixed various plants and dissolved them in a special wine called Kushumma, and added oil extract from cypress- three or other trees to the final product. Sometimes clay in powder form were added to the mixtures. Finally, they mixed water, honey and the water of the bottom of the sea to the final product.

For making medicine in the form of drops, minerals including sodium chloride or potassium nitrate were used. Animal products such as milk, snake skin and the shell of tortoise were popular. Generally, in Mesopotamia (Sumer, Babylonia, and Assyria) animal and mineral drugs were used less than herbals.

The Mesopotamian medical prescriptions indicate that they extracted medical decoction from tree bark; the solution of it was added minerals or alkaloids. Saltpeter was obtained by scraping the inner surface of sewer lines, where organic nitrous waste was deposited.

The Oldest Medical Documents

In the excavations of Nippur (modern Niffer, in Iraq), ancient city of central Babylonia (an important city as a Sumerian and Babylonian religious center), the oldest documents which are preserved in Pennsylvania Museum (in the USA) shows the methods of treatment of patient. One of the documents says: "After grinding the tree of white pear and the root of moon plant dissolve the mixture in beer, and administer the product to the patient. Grind the seed of the plant, harvest, the markasi resin and thymus in a container and dissolve the mixture in beer and then administer it to the patient.

Social Setting

From Middle Babylonian Nippur (ancient city of central Babylonia) there is various letters from physicians who practiced medicine. Among their patients, a group of them were females; probably they were musical students. In any case, it appears that they were housed as a group. It seems unlikely that this should be considered a hospital setting, however. These letters are addressed simply to "my lord," not otherwise identified, but surely a high official. An example of such a letter is given here in translation: "Say to my lord: thus (says)- muball it, your servant: I am ready to die as my lord's substitute. As to] about whom my lord wrote me, their are well. The patient whose chest is sick was prescribed a dressing, and is kept bandaged; he is also taking a potion against tracheitis. And the other one whose chest; when I assigned a poultice for him, no ašû herb was available, and my lord knows that if only a single herb is missing, it will not succeed. I asked the mayor to send word to a gardener. The daughter of Muštālu who was coughing but not spitting out has started to expectorate after I gave her [appropriate] potions; but now she is constipated. I gave her a potion for constipation to drink, and she is taking it, (but) there is no šarmadu herb and drawn wine available. Let my lord send (some) so that I can have her drink, lest she develop "Hand of Curse." The princess who had repeated attacks of fever has now calmed thanks to the dressing and potion. As to the herbs of which I spoke to my lord, let my lord not forget about them.

There are a number of letters from the court physicians dating to the time of the Assyrian King Esarhaddon. In these letters there are not only herbal medicines being prescribed (a practice we also saw at Mari in the letter cited above), but the use of magical treatments as well. Esarhaddon's illnesses have been studied in considerable detail.

Surgery

About three thousand years before the birth of Christ the Egyptians prepared the earliest known treatise on surgery. For the next 4000 years surgery progressed slowly impeded by beliefs in magic and devils as sources of illnesses and by unwillingness to dissect the human body after death to discover the cause of disease and to obtain accurate anatomical knowledge.

There has been discussions over many years about the extent to which physicians in Mesopotamia practiced surgery. Such discussions most often concern the eye operation mentioned in the Code of Hammurabi, sections 215-220. Whatever the procedure may have been, it is unlikely that any very complicated surgery was undertaken. The physician's scalpel was most likely used mainly to lance boils and for bloodletting- the latter sparsely attested in the medical texts. It is not known whether bloodletting was a widespread practice.

Very little paleopathological study has been published on human remains from ancient Mesopotamia for, unfortunately, it used to be that human bones were rarely preserved by archaeologists excavating sites of the historical periods. Probably the most extensive collection is from the excavations of J. N. Postgate at Abu Salabikh from the third millennium. Some of the graves have been published, but detailed paleopathological studies have not yet appeared.

Trephining

It is the oldest- known surgical procedure. Skulls of ancient peoples from thousands of years ago have been found with trephine (a small drill or bit with central points, which acts as a pivot) holes. The procedure was probably used for supernatural reason, but may also have been used to relieve increased intracranial pressure, as it is today. But there is no evidence in Mesopotamia for trephination (drill to bore a hole through the bony table, or layers, of the skull, and the use of such a drill), though it appears that it was practiced at least occasionally in other parts of the Near East, especially Egypt. As far as we know, circumcision was not practiced in Mesopotamia, though, of course, it was widely practiced in other areas of the Near East, including Jericho (modern Ariha, Jordan), ancient city in the Jordan valley of Palestine.

Human Anatomy

An early interest in the study of the structure of the human body and animals is reflected in the cave drawings of prehistoric times, the clay representations of human figures from the pre- Greek Minoan civilization of the Mediterranean, and models of supposedly prophetic sheep livers from Mesopotamia. The Egyptian gained some anatomical knowledge from the practice of embalming the dead, although this same practice apparently prevented dissection of human bodies. Egyptian papyri and account of surgery from Mesopotamia also indicate some anatomical sophistication.

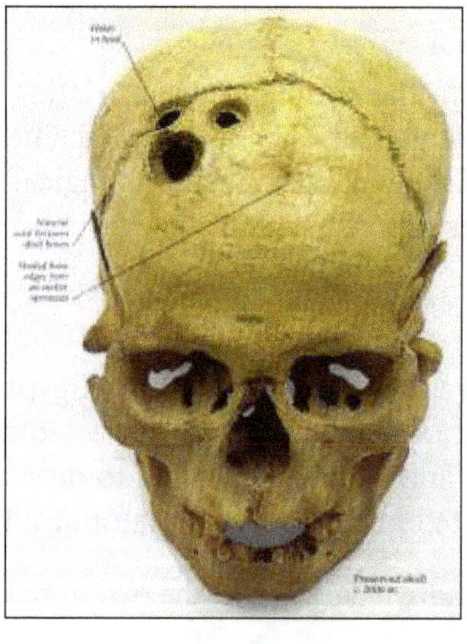

This skull which comes from Jerico, is 4000 years old, has three trephined holes. Drilled holes were roughly circular, whereas knife- cut ones were usually more square. A few skulls have up five holes- the largest ones measuring more than 5 cm (2 ins) across. Some people, including this individual, survived the process- this can be deduced from the fact that the bone signs of healing.

Knowledge of human anatomy was probably rather limited in Mesopotamia because, dissection was not performed. Nevertheless, it seems likely that there was some understanding of the body parts (if not necessarily their true functions) by analogy with slaughtered animals, particularly sheep, for which there was very detailed knowledge because of the practice of inspecting the internal organs as form of divination. Some knowledge of internal organs may have been acquired by physicians when they attempted to treat patients who had been gored by a bull or wild pig, or had been wounded in battle.

Without knowledge of the true etiology of specific diseases, physicians could treat only their overt symptoms. It is recognized that in traditional medical systems the practitioner need not necessarily know the cause of an illness to deal effectively with it. It is also a fact that most diseases (degenerative diseases excepted) are self-limiting, and the patient either recovers or dies.

Childbirth and Caesarian

Childbirth, process of giving birth to a child, occurs about 280 days after last menstrual period, unless the child is delivered prematurely. Various complications may develop when the child passes through the narrow opening of the pelvis. If the child is too large for the pelvic opening, delivery may have to be accomplished by Caesarean section. The term Caesarean section though probably derived the Latin word "caedere" ("to cut"), is popularly thought to be named after Julius Caesar (102-44 BC, Roman statesman and general, and one of the greatest figures of antiquity), who was said to be so large at birth that a normal delivery was impossible. Caesarean section was performed with close attention to technique in order to save both mother and infant. For a dead child still in the womb, dismemberment through the vagina was painstakingly carried out to avoid the risks to mother of opening her abdomen. Recognizing the hazard of abnormal presentations at delivery, writers gave instructions on how to turn a child into the proper position.

Oppenheim believed he had found evidence for the cesarean section in Babylonia his proposal has often been taken as fact. Despite the addition of further texts and continued discussions by several scholars, it must be said that a convincing case has not been made and that it remains an unproved suggestion. But female problems connected with pregnancy and childbirth are often addressed in the Mesopotamian medical texts. There are many prescriptions for a physician in the medical texts that physicians helped the delivery of babies. Rather, the assistance was provided by a midwife, probably aided by a female relative very many texts, however, offer prescriptions for giving birth easily and for making a barren woman conceive. An instance is "Total: 21 stones to help barren woman become pregnant; you string them on a linen thread and put them around her neck."

There is a text that gives prescriptions for abortion. The relevant line reads, "to cause a pregnant woman to drop her fetus." The prescriptions include of eight ingredients to be administered to the woman in wine and to be drunk on an empty stomach. The section ends with the sentence, "that woman will drop her fetus."

Illnesses of infants and children are detailed in chapters of the omen collection known as SA.GIG, but children's illnesses are not mentioned in the therapeutic medical texts. From the Mari letters that physicians did in fact treat children. Children's illnesses are also addressed in the magical series Lamaštu.

Leprosy

Leprosy or Hansen's disease is a chronic infectious disease which involves primarily the skin and the peripheral nerves, would not at present be justified in suggesting its occurrence in ancient Mesopotamia at an earlier time. But evidence now suggests it occurrence in Palestine and the Byzantine period. Plagues and epidemics were mentioned often in the Bible (sacred writings of the Christian church, comprising the Old and New Testaments), with special attention given to leprosy, which was feared and isolated, but, as among the AssyroBabylonians, many skin diseases considered to be leprosy probably were not. There were, however, references to many other types of illnesses and symptoms in the Bible. The precepts and prohibitions of Biblical times on personal and public health were continued in the Talmud (collection of ancient writing on Jewish law and tradition) and the leper was regarded unclean and his or her cloths were to be burned.

Pharmacopeia

Pharmacopeia, also spelled pharmacopoeia, a book containing a list of medicinal substances with descriptions, formulas, properties, identification tests, quantitative assays, safe dosages, and other standards to determine their strength and purity. Although descriptive lists of drugs are known from ancient China, Mesopotamia (including summer, Babylonia, and Assyria), and Egypt, the precursor of the Pharmacopeia is generally considered to be a book entitled Composition, written by the Roman physician Scribnius Largus about 43 AD. The "Nuovo receptario composito," published by the Florentine guild of physicians and apothecaries in 1498, was made official and thus became the first pharmacopeia of the Europe.

In Mesopotamia, about three to four thousand years BC "most of the medical treatments" consisted of herbal remedies. As in other traditional societies, the most common purposes for which herbal medicines were used were to treat gastrointestinal, dermatological, and respiratory ailments. Some of these remedies quite likely produced the desired effect, especially in the case of emetics, purgatives, and expectorants. The beneficial effects of some common products (such as wine and honey) are well known and have been demonstrated in modern experiments. It is well known that certain herbs can produce vomiting (in fact, emetics are often toxic to some degree, which is why the body's natural defense system causes them to be expelled) and that others are effective purgatives. A good many herbal remedies were probably pharmacologically ineffective, however, with benefits largely limited to calming a patient's anxiety.

The Babylonian medical texts prescribe a great many plant products, usually specific parts of the plant such as leaves, blossoms, seeds, or roots to be prepared in various ways (crushed, cooked, or dried, for example) and mixed with an appropriate carrier (water, beer, wine, or milk for a potion, for example). Knowledge of plant chemistry as such in traditional societies, there must have been some knowledge of toxicity in plants, for, indeed, only specific parts of many plants are toxic or are toxic only at a particular stage of their grow cycle. In fact, many of the plants used in medicine are normally not consumed as food, and thus presumably had some disadvantage such as toxicity or unpalatability. Of course, in using many of these drugs, care had to be taken to use a concentration that was therapeutically effective but nontoxic. Many prescriptions call for a large number of ingredients, often combining plant products and minerals. In many cases, the ingredients are summarized as, for example, "these fourteen ingredients you take and prepare.

Dentistry

In Mesopotamia, it was believed that demons inhabiting in the body were responsible for illness. Filthy and disgusting remedies concocted from worms and insects were used to sicken and drive out these evil spirits. The Code of Laws now preserved in the Musée de Louvre, clearly spells out the punishments and rewards due practitioners of medicine:

- Law 196: If someone injures the eye of an equal his own eye is destroyed.
- Law 198: If someone injures the eye of an inferior he is fined a mina of silver.
- Law 200: If someone knocks out the tooth of an equal his own tooth is knocked out.
- Law 201: If someone knocks out the tooth of an inferior he is fined a third of a mina of silver.

It is interesting to note that, although an eye was clearly considered more valuable, nevertheless a substantial sum was placed on the value of a tooth. Among the clay tablets in Ashurbanipal's library are a number devoted exclusively to diagnosis and prognosis, and the state of the teeth was used as a means of determining the course and source of an illness:

> "If he grinds his teeth the disease will last a long time. If he grinds his teeth continuously and his face is cold he has contracted a disease through the hand of the Goddess Ishtar."

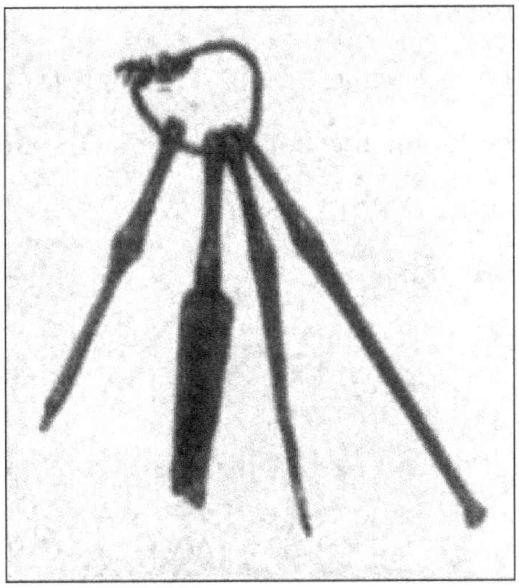

At the site of Tepe Gawra, about twenty miles from ancient Nineveh, archaeologists found this four- thousand- year-old set of toilet articles, which includes a makeup applicator, an ear scoop, and a finely crafted toothpick.

Since grinding the teeth was considered both very dangerous and pathognomonic, a remedy was proposed. A human skull was placed upon a chair, and for three days, morning and night, sacrifices were placed before it. Then conjurations were to be spoken seven times into the skull, and the skull was kissed seven times seven by the patient before retiring, and then he would become well. The royal libraries yielded an intriguing letter written by the court physician of the Assyrian king Essarhaddon which reads, in part, "As regards the cure of the [aching] teeth about which the king wrote to me, I will [now] begin with it; there is a great lot of remedies for [aching] teeth." Unfortunately, the letter does not tell us what those remedies were. Another letter, in response to a king's inquiry about his son's illness, states, "The inflammation where with his head, hands, [and] feet are inflamed is due to his teeth. His teeth must be drawn then he will be well.

This belief that a tooth worm causes dental caries (decay in teeth) was first documented in Babylonia. A clay tablet found in royal library gives us a detailed account of dental caries with poetic starkness:

- After Anu [had created the sky].
- The sky had created [the earth].
- The earth had created the rivers.
- The rivers had created canals.
- The canals had created the marsh.
- The marsh created the worm.
- The worm went forth weeping, before Shamash.
- Before Ea in tears (saying).
- "What will you give me to eat?
- What will you give me to suck on?"
- "I will give you ripe figs, armanu fruit, and apples?
- Of what use to me are ripe figs armanu fruit, and apples?
- (Instead), raise me up and let me live between the teeth and the jaw.
- I will suck the blood from the teeth.
- I will chew upon the food in the jaw."
- (Instructions to dentist) Drive in a pin and seize its foot.
- "Because you have spoken thus, O worm.
- May Ea strike you with all the strength of his hand."
- Text for tooth trouble.
- Procedure: you mix together beer, a lump of malt, and oil.
- You recite the (above) incantation over it three times.
- You put it on his teeth.
- The colophon, indicating that this is an old text, reads as follows: "Copy of a long tablet,

according to the wording of an old baked tablet belonging to Marduk- nadin- akhi; Nabun-adin- ipri, descendant of Kudurranu, copied it."

There were also barbers who treated tooth disorders and did extractions.

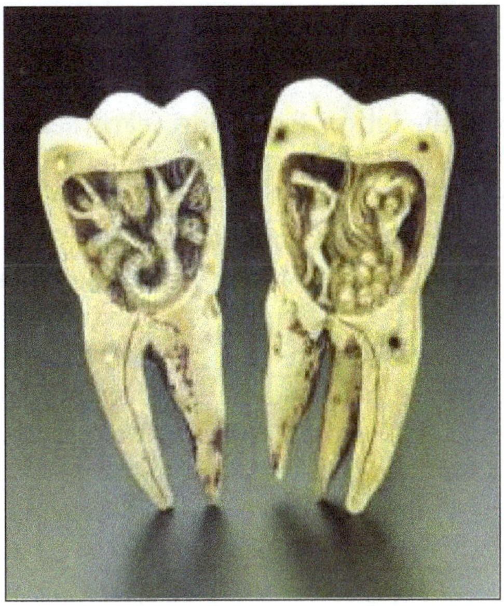

The belief that tooth worm causes dental caries stubbornly clung to until the 18th century. An artist in southern France carved an ivory replica of a human molar tooth, about four inches high, which can be opened to reveal on the left a tooth worm devouring a man. On the right, the torment of toothache is equated with the tortures of hell.

Public Health and Hygiene

Today, in highly industrialized part of world each citizen is enveloped in a network of vital public health services (the organized efforts of the community to protect its members against diseases), which, although inconspicuous and often unappreciated, have helped to extend his average length of life about 20 to 30 years within the span of over a century. But circa 4000 BC in Mesopotamia and other parts of the world, water was not purified of typhoid and cholera bacteria and other disease- producing microorganisms. The milk was not pasteurized to destroy the germs of tuberculosis and undulant fever. Early in life, the Mesopotamians were not immunized against smallpox, diphtheria, whooping cough, plague, and other communicable diseases. There were no sewage- disposal systems, supervised by public health agencies to prevent the human wastes from becoming a source of infection. Therefore, epidemics must have occurred often; the many wars and invasions were likely to foster pestilence. Frequently, plagues of some kind were reported in the cuneiform tablets of the eighth century B.C., and fevers, probably of varying causes, were mentioned frequently in the medical texts. The shaking chils which Alexander the Great suffered in his last illness while campaigning in Mesopotamia in the fourth century B.C. may have been due to malaria.

A sick person of any rank in Mesopotamia was in a special category and was excused from work and even from service to the king. On the other hand, since disease was caused by spirits having possessed the body, the afflicted person was shunned as much as possible to avoid transference of the offending demon. This relative isolation was hygienically beneficial to the community although its purpose was based on religious- magical reasoning. The taboo against touching the sick was carried over into Hebrew culture, where it became a key factor in a system of public hygiene- just one further example of Mesopotamia's long- lasting influence on contemporary and later cultures. For instance, one of the earliest public health programs is detailed in the Biblical account of the Exodus of the Jews from Egypt. Moses introduced a strict hygienic code which included many principles observed in modern public health programs. The Jews were required to wash their hands before and after eating and to bathe before entering the synagogue.

In Mesopotamia, it is clear from the letters from Mari (early second millennium) that measures were taken to try to control the spread of apparently contagious illnesses among the population. A number of letters refer to populations of whole villages being moved elsewhere, normally to higher ground. There was indeed a realization that some illnesses could be transmitted easily from one person to another, so, at least to some degree, there was isolation of the ill when their illnesses appeared to threaten the community. The following letter illustrates such a situation: "I have heard that the woman Nanna is ill with simmum disease, but she has (nevertheless) been in contact a great deal with the palace (servant women) and that she has infected many of the women around her. Give strict orders that no one drink from a cup she drinks from, that no one sit on a chair she sits on, and that no one sleep in a bed she sleeps in so that she does not infect (any more) of the many women around her. That simmum disease is (easily) caught!" Indeed, the chances of encountering infectious microorganisms or viruses must have been quite high in the close quarters of palace life.

Some illnesses which gave every appearance of being contagious were surely mycotoxicosesillnesses caused by consumption of cereal crops contaminated by toxic molds (the toxins in these molds are unaffected by cooking and baking). The evidence is necessarily circumstantial, but it is surely significant that some texts mention illness affecting both humans and domestic animals, especially cattle. It is known that the livers of cattle are less able to process toxins than are the livers of sheep, and therefore it is not surprising that the texts specifically mention illnesses of cattle. Descriptions of the condition of sick bulls in Mari letters are consistent with modern evidence concerning cattle poisoned by toxic molds, particularly when the poisons result in gangrene of the extremities.

Temple of Gula

Beyond the medical practitioners (ashipu and asu), there were other means of procuring health care. One of these alternative sources, the Temple of Gula which often envisioned in canine form, was one of the more significant gods of healing. While excavations of temples dedicated to Gula have not revealed signs that patients were housed at the temple while they were treated; these temples may have been sites for the diagnosis of illness. In his book Illness and Health Care in the Ancient Near East: the Role of the Temple in Greece, Mesopotamia, and Israel, Hector Avalos states that not only were the temples of Gula sites for the diagnosis of illness, but that these temples were also libraries that held many useful medical texts.

The primary center for health care was the home, as it was when the ashipu or asu were employed. The majority of health care was provided at the patient's own house, with the family acting as care givers in whatever capacity their lay knowledge afforded them. Outside of the home, other important sites for religious healing were nearby rivers. The Mesopotamian believed that the rivers had the power to carry away evil substances and forces that were causing the illness. Sometimes as small hut was set up for the afflicted person either near the home or the river to aid in the families centralization of home.

Medicine was one of the sciences that was founded by priests of Babylonia. They were the first people of the world who studied the knowledge of medical treatment. The methods they used were unique. They left patients on foot- pavement and roads, and those passer-bys who were inflicted with similar ailment and had succeeded in treating their illness, would voluntarily reveal their information. Chaldean priests would write such information on clay plates and hang them in the temples for the benefit of people. Obviously, they practiced medicine in this way. Other nations including Arabs and Greeks followed Mesopotamian's medicine (Temples known as "asklepia" were built to Asclepius, god of medicine, throughout the Greek world and sufferers spent the night at the temple and the next day described their dreams or visions of him to the temple priests, who gave them a advice.

Herdotus Greek historian, called the father of history whose travels took him to the coast of Asia Minor to the shores of the Black Sea and Skythia, to Phoenicia, Mesopotamia, and Egypt states: "They do not have physician, therefore, they leave their patients on side- walks and streets. When a passerby who has proper medical knowledge on the basis of personal experience or visiting similar cases, encounter these patients, he may give advice. Usually the passerby would stop beside the patient and advise him on the treatment that was successfully administered to him or someone he knew. Passing by the patient without stopping or being indifferent about the situation was not ethical and everybody had to converse with the patient and ask about the patient's illness."

Beginning of Public Health Mentality

In Assyria, another feature of their medicine was the beginning of a public health mentality, produced by a strict code of religious ceremonial laws. A woman after childbirth was considered unclean and contaminated everything she touched. A similar impurity was attached to those who came in contact with corpses. These ideas are reflected in the earliest Persian medical works which we know, the surviving books of Zoroaster, and it is clear that the Persians, even before they conquered the Assyrians in 538 BC, had largely adopted the medical and sanitary ideas of their neighbors on the plains.

An Ideal in Babylonia

To enjoy good health and to have contentment, to be viewed favorably by the gods and fellow humans and to enjoy a long life with one's descendants were the ideal of Babylonians. A long life included having children who would provide funerary offerings after one's death. A statement attributed to Adad- guppi to female diviner and the mother of the Babylonian last king Nabonidus telling of he long and contented life (which she attributed in part to her piety and zealous worship of the moon- god) illustrates the point: "He (the moongod) added many days and years of happiness (to my life) and kept me alive one hundred and four happy years. My eyesight was good (to the end), my hearing excellent, my hands and feet were sound, my words well chosen, food and drink agreed with me, my health was fine and my mind happy. I saw my great- great grandchildren, up to the fourth generation, in good health and (thus) had my fill of old age.

This quotation gives an indication of what approximated an ideal. In real life, however, there were many deviations from this ideal, and prayers seeking deliverance from troubles often enumerate some of them: crying spells, insomnia, nightmares, fears, inability to enjoy food or drink, sexual inadequacies, and so on.

Generally a person who considered himself ill was considered so by those around him. The symptoms described usually include those perceived by the patient, such as localized pain or itching, nausea, or the inability to get up and walk. In other cases, it is those around the ill person who perceive him to be ill, as when his behavior is irrational, he keeps forgetting what he has said, or he does not recognize people he knows.

It is probable that home remedies were administered in many cases, and these were most likely fairly simple traditional treatments, utilizing medications that did not carry any unusual risks. The various apothecary's inventories that have been preserved may indicate, however, that a good many of the most commonly utilized ingredients were readily available to whoever sought them. Whether or not a professional healer had prescribed them for a particular illness.

Urban Morphology

Urban geographers have made important contributions in the field of spatial transformations that have been witnessed by urban landscapes in the material and symbolic aspects during the twentieth and twenty-first centuries. Some of the landmarked attempts in this direction were the analyses of urban morphology done by Burgess, Hoyt, Harris and Ullman. Although cities at present have changed significantly since the models were developed; yet they are frequently cited in debates of urban morphology even if to dismiss their continued relevance. That is very true say that a large extent of each city possesses a distinctive combination of varying types of land uses, but a common pattern is can be traced some time. The models provided by Burgess, Hoyt, Harris and Ullman, today are part of the philosophy of urban geography and one needs to discuss them in order to understand the basic foundations of this field.

The morphological pattern of a city has been discussed by three theoretical explanations. These are:

- The Concentric Zone Model.
- The Sector Model.
- The Multiple Nuclei Model.

The Concentric Zone Model

The concentric zone model was given by E.W. Burgess in 1923 based on his extensive and detailed case studies on Chicago and its various neighbourhoods. His model is radiating out from the Central Business District and represents increasing degrees of cultural assimilation with greater economic and social status through each successive residential zone. The zonal model became a significant and long standing representation of the North American city especially to study the correlation of social and spatial distance among the various classes of the society.

Before understanding the Burgess's model of Concentric Zone one has to study the tendency of the city to expand. The typical processes of expansion of the city can be best demonstrated by a series of numbered concentric circles as shown in Figure. In the figure, Zone I 'the Loop' represents the Central Business District (CBD). Encircling this is an area in transition which is usually surrounded by business and light manufacture Zone II. The Zone III is inhabited by the workers in industries who have escaped from the area of deterioration (Zone II) and have the desire to live within easy access of their work. Next is the residential area Zone IV of high-class apartment buildings or a district of single family dwellings. Zone V is beyond the city limit – the commuter zone comprising of satellite cities or suburban areas; within a thirty to sixty-minute ride of the CBD. And each inner zone has the tendency to extend its area by the invasion of the next outer zone. This basically deals with the physical growth of the city and with the extension of the technical services that make city life liveable, comfortable and luxurious.

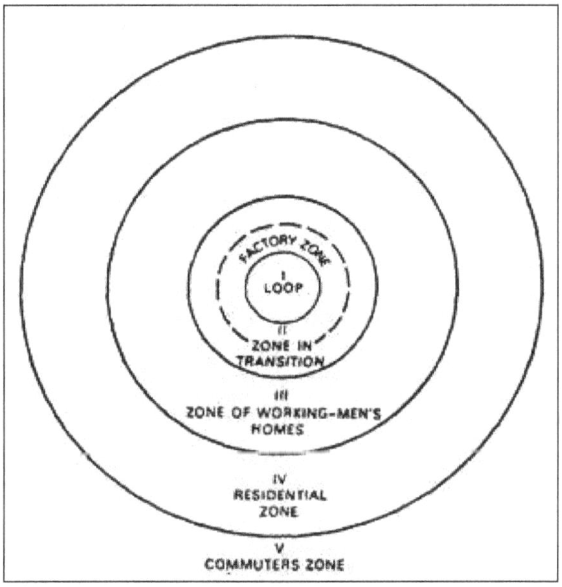

The Growth of the City.

Burgess through his model presented an image of the progressive movement as residents of the inner city had a tendency of moving outward to zones of better environmental conditions. The American city should take the form of five zones. These zones are:

Zone I- Central Business District (CBD): It represents the area of CBD revolves around social, commercial and civic life. It is also the hub of transportation. While Burgess explains its two parts:

- The downtown retail district.
- The wholesale business district encircling the downtown.

Zone II- Zone in Transition: It is an area of which surrounding the Central Business District (CBD): It is a zone of residential deterioration of old private houses consisting of largely sub-divided dwelling units. The transition area is settled by immigrants and overrun by 'vices'. The Zone II full of rotten business and light manufacturing units from Zone I encroach upon residential belts. Some portion of this zone is likely to be found in the city's slums or poverty areas and crime. This is a zone with mixture regions of poverty, disease and their criminal activities and vice.

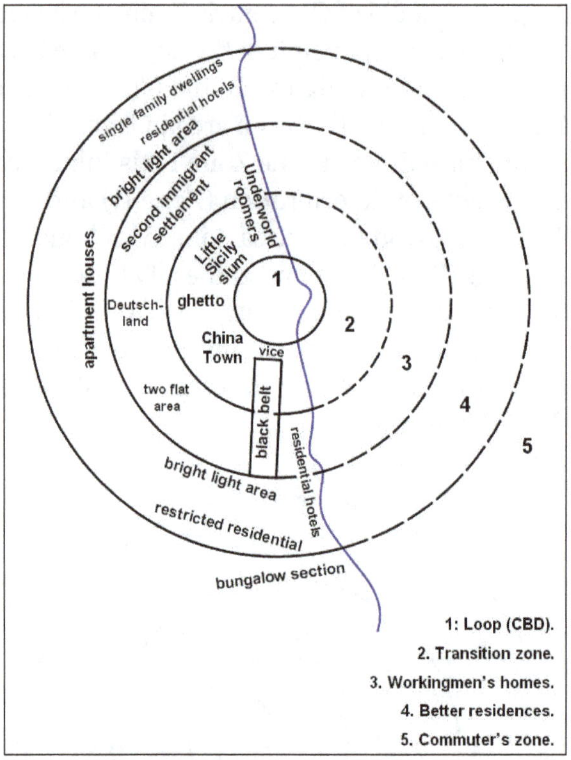

Concentric Zone Theory.

Zone III- Working Men's Homes: It is the third ring sum up with the houses of working people's homes. This is predominately inhabited by factory and shop workers who are skilled and thrifty. It is the zone of old residential blocks occupied by social groups of working class groups and families those are stable in their jobs. People of this zone moved out of Zone II to live nearer to their work place. This is an area of second immigrant settlement, generally of the second generation. It is the region of escape from the slum or over crowed places.

Zone IV: It is the zone with concentric space still farther from the CBD and is settled by spacious dwellers. In Chicago, this was dominated by native-born Americans of middleincome groups. The population residing in this zone are likely to be small proprietors businessmen, salesmen, professionals and different official clerks.

Zone V: It is the areas farthest from the city centre; and almost one hour's to two hour's travelling time from centre to this zone. This zone may be an open country or country side region. People of this region seem to commute on a daily basis for their livelihood in the centre.

Criticism

With the few modifications Burgess,' theory is popularly and widely used by urban geographers. But the Burgess' theory is criticized on the several grounds like local topographical features which affect the residential areas location. While this type of criticism looks invalid because Burgess himself criticised that point i.e. those zone distortions may result from variations in relief features. Different researchers have criticised the theory on following grounds:

- The size of CBD's is irregular in the pattern; and is often rectangular than circular.

- Commerce and business areas of usually extend along streets from the CBD's in a radial way.
- Along the lines of transportation, near water or rail network Industrial units lie.
- Near to industrial and transportation areas in every zone the houses are low grade generally.
- Lastly, Burgess' theory lacks universal acceptance as a whole.

The critics of Burgess's concentric zone theory focus on that the theory is not applicable in the case of its treatment of wholesaling market. Similarly, in the modern city the large and heavy industries do not take the form of the concentric belt just outside the CBD, instead, it forms wedges like areas along transportation lines.

The theory of Burgess seems weak, in the historical perspective too. The houses, streets, and railroads of culture areas developed during historical phases very hard to change their location. The theory was considered generally in time and space, and it was outdated and limited only to large Western industrial cities by the late 20th century.

The concentric scheme for the 'pre-industrial city' neglected by researchers in which privileged classes – the elite, gather at the centre due to the nearness of governmental and religious buildings. While religion and politics in feudal cities had far more important than the economic –in which the centre's main market being subsidiary to religious and political structures.

Merits of the Concentric Theory

The chief supporter of the Burgess theory says common-sense observations which tend to the theory confirmation. The probability of the concentric structure around the dominant retail area in various cities indicates by the Urban-gradients' researchers.

The symmetry of concentric may be violated by Local irregularities; the most cities conformed at least roughly to the Burgess pattern. The contribution of Burgess's model too had appreciation by various researchers according to them it was a normative model, 'which is based on simple structure of reality presenting supposedly significant features of relationships in a generalized form. To conclude, the illustration of burgess's model of concentric zone theory; is an expansion of the city in its comprehensible manner by a series of concentric zones.

The Sector Theory

Burgess's concentric zone theory was based on the American cities morphology in 1920s. Homer Hoyt observed that the cities structure had changed and therefore, Hoyt propounded the sectoral theory in 1939. He emphasizes on the ribbons of development that extended along commercial streets leading out of the central business district and the tendency of industrial establishments to concentrate along railroad lines and rivers. He further notes the existence of poor and middle-income housing along the rich residential areas in the peripheries of the cities. He opined that the internal structure of the American cities was more axial than concentric resulting in the formation of different sectors. As the city grows, activities expand outward in a wedge or an axis or a sector from the centre.

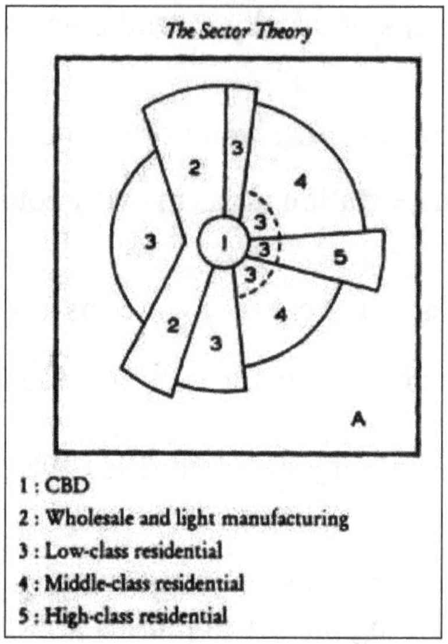

Hoyt's sector theory, which explain only with residential land use pattern; the other types of land uses are considered because of their influence upon the residential areas of the city. The cities areas which are on rent tend to more and less sectors rather than of concentric zones. The areas located in one or more sectors of the city tend to be the highest rent areas. But it is not necessary that all areas are rent areas. The migration of outward sector is clearly visible in high quality residential areas; where the older settlement remains behind to become medium-quality residential areas. Hoyt stated that the sector theory is of fundamental importance in analysing neighbourhoods especially in the American cities for locating markets for retail sales. The high rent neighbourhoods of a city do not skip about at random in the process of movement-they follow a definite path in one or more sectors of the city. No city conforms to the ideal pattern but the general figure is useful as in American cities the different types of residential areas tend to grow outward along rather distinct radii and the new growth on the arc of a given sector tends to take on the character of the initial growth of that particular sector.

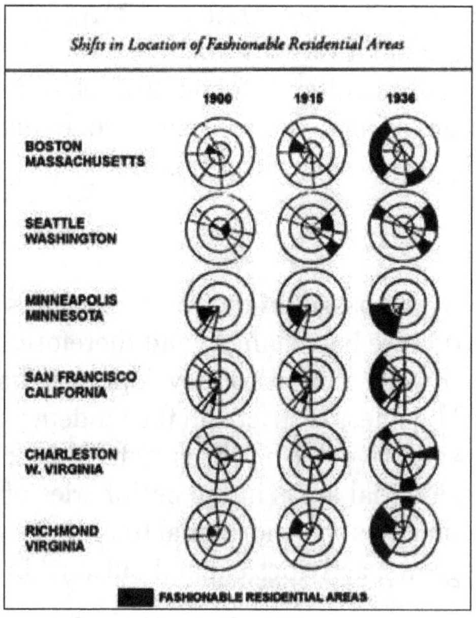

The basis of Hoyt's theory is a bundle of empirical work. The 64 American cities were the sample size of Hoyt's theory which is collected by the Works Progress Administration. The surveys of New York, Detroit, Chicago, Washington and Philadelphia supplemented the small and medium sized cities. Thus, to empirical generalizations it is a very bulk part of observation. Nevertheless, the theory has not gone unchallenged.

Criticism of Hoyt's Theory

The theory of Walter Firey's carried out a land use study of central Boston. In which he explored the urban land use shaping by the role of social factors. He pointed out various aspects of sector theory on this basis of his survey. He argues that there is less validity of comparing the internal structure of number of cities when physical features like relief, location on a waterfront and other factors affect the pattern of some cities. Firey also criticised that Hoyt theory has not sufficiently considered the roles of social and cultural characteristics in land use explanation. It is obvious that richer residents can choose to settle anywhere and may not follow the 'normal' pattern i.e.; either concentric or sector pattern.

The Multiple Nuclei Theory

Harris and Ullman brought together their work on central palace theory and classification of cities respectively to represent a metropolitan area that was not defined by distances from the central business district but was based on patterns of land-use of the surrounding areas of the CBD. During the time of its inception to the present day, taking place in cities, residential, business, industrial or other elements to accommodate the growth of multiple nuclei model suggested by them.

Harris and Ullman unlike Burgess and Hoyt stressed upon the development of special purpose districts instead of social differentiation of residential areas. These districts were nodes of economic activity that needed specialized facilities and benefitted from agglomeration economies of different activities centred there. The rise of such special districts or nuclei was due to numbers of factors which affect the human activities expansion within a city. These are as fallow:

- Some activities require specialized facilities like, CBD; which can function at the point where maximum accessibility possible.
- There are also numbers of activities which is present in cohesion. The densely packed inner districts of large cities are clustered with clothing industry. Due to cohesion they get profit.
- Some activities are detrimental in nature to one another, and generally seek separate locations. For example; heavy industry does not prefer to be nearer to the high-class residential areas.
- While some are unable to pay the rents of most desirable locations; like low income residential areas or bulk storage facilities have to seek nuclei in remote regions.

The factors along with social, cultural and economic criteria provide a unique urban landscape with separate nodes. Moreover, this theory reveals two significant observations based on historical and site elements of morphology. In one hand the theory produces a model involving complexities of urban structure which may not be easily and immediately discernible because of historical stratification of land uses during the process of urban growth. Although there is only one CBD present in most cities and they have a series of sub-centres around CBD. The peripheral areas are less specialized than CBD but enough to cope up the needs of smaller sections of the city.

Multiple Nuclei Theory

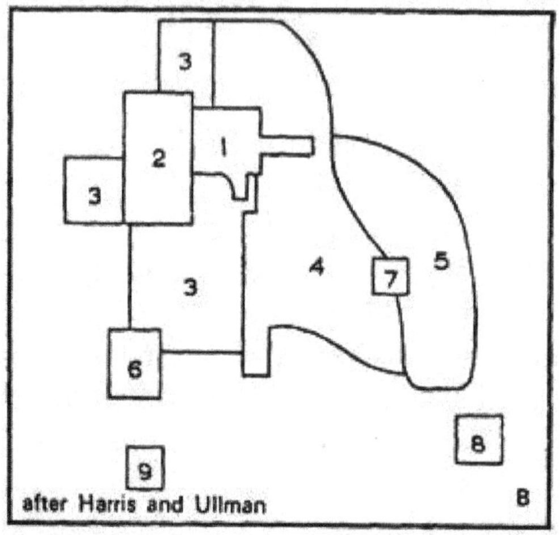

Central Business District	1	Heavy Manufacturing	6
Wholesale Light Manufacturing	2	Outlying Business District	7
Low-class Residential	3	Residential Suburb	8
Medium-class Residential	4	Industrial Suburb	9
High-class Residential	5		

Another one is an observation which is more significant is about the probability of elements of the concentric and sector models present in its depth. There is nothing new is involved conceptually in the multiple nuclei model, and, it should not be given the status of a theory. Therefore, multiple nuclei theory should be looked upon as an approach which only road to think about the structure of the city, rather than as a rigid generalization in nature about urban form.

Urban Poverty

If poverty means human needs that are not met, then most of the estimates for the scale of urban poverty in Africa, Asia, Latin America and the Caribbean appear too low' For instance, a publication by the Overseas Development Council in the USA in 1989 decided that only 130 million of the Third World's 'poorest poor' lived in urban areas which meant that more than nine out of ten of the Third World's urban population were not among the poorest poor. World Bank estimates for 1988 suggested that there were 330 million 'poor' people living in urban areas which meant that more than three-quarters of the Third World's urban population were not 'poor' on that date. These figures do not seem to fit with national and city studies which show that one-third to one-half of a nation's urban population or a city's population have incomes too low to allow them to meet their needs. National studies in several of the poorest African, Asian and Latin American countries suggest that more than half the urban population are below the poverty line. In addition, many studies have shown how, in the 1980s, urban poverty increased considerably in nations experiencing poor economic performance and/or structural adjustment.

It is not only international statistics that seem to underestimate the proportion of urban households living in poverty but also many national statistics. In most countries in the South, the scale of urban poverty is measured by the number of households whose income falls below a government defined 'poverty line' - and many such poverty lines are set at income levels that appear unrealistically low in relation to living costs. Thus, in many countries, a considerable proportion of the urban population who, according to official statistics, are 'above the poverty line' (because they have above poverty line incomes) still lack the income they need to cover the costs of basic necessities - which include safe and adequate quality housing with piped water and adequate provision for sanitation, and expenditure on health-care and on supporting children at school.

One reason why income based poverty lines are set too low in relation to living costs for many urban households is the assumption that living costs are the same, wherever the household lives. In many countries, a single income level is set as the 'poverty line' for both rural and urban households. This assumes that the costs of the basic necessities that must be paid for to avoid poverty are the same for rural and urban households. But the costs of these necessities are usually higher in urban areas than in rural areas, for example, the cost of building or renting housing, getting to and from work and paying for water. In most urban areas (especially larger ones), there are also fewer opportunities for reducing costs through some subsistence production (e.g. growing food) or through access to free resources (e.g. wood for building or for fuel)? The cost of basic necessities is likely to be greatest in the larger and/or more prosperous cities. It has become more common for governments to set separate poverty lines for urban and rural areas on the assumption that living costs are higher in urban areas; but most urban poverty lines still seem too low in relation to the cost of basic necessities.

Many poverty lines that are applied to measuring the scale of urban poverty assume that it is a household's capacity to purchase (or obtain) food that is the main determinant of whether they are poor. Thus, the income level at which the poverty lines are set is much influenced by the cost of food. Even where poverty lines are based on a 'minimum consumption basket', it is usually the cost of food items that take up much of this 'basket'. But avoiding poverty in urban areas also means having access to income-earning opportunities and, for many urban households, the cost of public transport getting to and from work is high. For some, it is even comparable to, or greater than, the cost of food. Avoiding poverty in urban areas also means access to safe and secure accommodation with basic services - and in many cities, the cost of the cheapest form of safe, secure and minimum quality accommodation with adequate provision for water and sanitation is likely to be considerably higher than the entire poverty line income. Of course, this is the reason why such a high proportion of the urban population in the South live in such poor quality, overcrowded and insecure accommodation with inadequate provision for basic services - because it is cheaper and is all they can afford. One study of poverty in Latin America, which took into account the real cost of living in different countries and which made allowances for the higher cost of living in urban areas, suggested that a much higher proportion of the urban population were poor compared to figures arrived at through the conventional measures of the World Bank.

The Income Needed to Avoid Poverty

One of the great unknowns is by how much the level of income needed by a household to avoid poverty varies from place to place. A single poverty line applied to a whole country assumes that there is no variation, which is ridiculous. Certainly, many (or most?) urban households need a higher cash income than many (or most?) rural households to avoid poverty:

- For public transport (for getting to and from work and essential services).
- For schools (where school fees and associated costs, including getting to and from school, are higher than in rural areas).
- For housing (for rent or, if living in a self-built house, because access to a land site for the house and building materials is more expensive).
- For access to water, sanitation and garbage collection (for many urban households, the payments made to water vendors represents a major item of household expenditure; many urban households also have to pay for garbage collection whilst some have to pay for access to latrines).
- Perhaps for food as food is more expensive (especially for urban households who have no possibility of producing any of their own food).
- Perhaps on health-care if this is more expensive in urban areas (or no public or NGO provision is available and private services have to be purchased).
- On child-care (where all adult members have to find income-earning opportunities and child-care is needed but there are no low- or no-cost solutions).

It is impossible to define a single income level which constitutes what all urban households need to avoid poverty because the income level needed varies considerably between urban centres. For instance, the costs of public transport, schools, health-care, housing, basic services, etc. are likely to be higher in the larger and more prosperous cities than in smaller and/or less prosperous urban centres. It is also clear that the income level that an urban household requires to avoid poverty varies considerably within cities. For instance low-income groups who have to live near the city centre because of the ready access this provides to casual jobs or to jobs with long hours generally have to pay much more for adequate quality accommodation with basic services than those groups living in the predominantly low-income areas of the periphery However, for low-income households who choose to live in low-income peripheral areas because accommodation costs are lower, there is more space, and because there are greater possibilities of home-ownership, transport costs are one of the main items of household expenditure, especially if one or more income earner is commuting to a central location.

The Extent of 'Hidden Income' in Urban Areas

Most urban households need substantially higher cash incomes to avoid poverty than most rural households, especially if they live in the larger and/or more prosperous cities. But some (or most) low-income urban households may enjoy much better basic services than most low-income rural households - and this can be considered an important hidden income'. Efficient public, private or community action can lower the income that a household requires to 'avoid poverty' A well-managed system for piped water and provision for sanitation, drainage and garbage removal (generally more common in urban than rural areas) can greatly reduce the cost of finding adequate quality accommodation with basic services. Effective public or non-profit private provision for schools, health-care and child-care can also lower the income needed by households to avoid poverty An efficient public transport system can cut the costs of access to employment while well-managed housing credit schemes can cut the cost of access to adequate housing (and to acquiring housing which then becomes an important asset). There are also many examples of relatively low income urban households gaining access to land for housing at below market price - although this is now less common as even informal or illegal means of obtaining land for housing have become highly commercialised in many cities.

There are also the many urban households who have no access to safe and sufficient water supplies, no provision for sanitation and drainage nor-access to schools and health-care services even if they have incomes that are above the poverty line. As such, they suffer from forms of deprivation that are generally associated with poverty but it is not their low income which is the cause but the incapacity of public, private or non-profit institutions to ensure provision. Here, it is important to clarify the extent to which many forms of deprivation faced by poor households (and often many non-poor households) are more the result of weak, ineffective, unrepresentative or corrupt governments than of their income levels. As such, many forms of deprivation associated with poverty can be addressed by more competent and effective public or private institutions - and, in many urban areas, with little or no subsidy, there is great scope in most urban centres in the South for supporting improvements in housing and living conditions, and basic services among lower-income households at low per capita costs and with a considerable degree of cost recovery.

Comparisons of differentials in the cost of living or the income level needed to avoid poverty between urban and rural areas are fraught with problems. These comparisons have to recognise that the income level needed to avoid poverty will vary greatly between different urban centres and within urban centres (especially the largest cities) depending on where the person or household lives. There may be comparable variations in the income level needed to avoid poverty for rural households, especially between the rural areas with the most commercialised and least cornmercialised agriculture, forestry and land markets. There are also large variations in the extent to which efficient public, private and community action lowers the costs of avoiding poverty (or is able to do so) - and, in general, it is likely that low-income groups in urban areas (or in particular cities) benefit more than those in rural areas. It may be that a much higher proportion of urban dwellers have access to safe and sufficient water supplies, to schools and health-care services, so that rural-urban comparisons of costs miss differentials in their availability Then there is the fact that the income level needed by a household to avoid poverty also varies according to the household (for instance relating to how many dependents there are in the household) and according to time (with expenditures varying according to the age of the children) - and perhaps also according to the season as incomes andlor prices fluctuate. The more a househoıd income level determines their access to housing and basic services, the more vulnerable they are to becoming 'poor' when the number of income earners within the household is at its lowest and the number of dependents at its highest.

Integrating a Concern for Housing and Basic Services within Poverty

One reason for questioning the validity of international and national statistics on the extent of urban poverty is that there seem to be far more households 'living in poverty' than are considered to have incomes below the poverty line. Hundreds of millions of urban dwellers in the South who, according to global estimates of poverty have incomes that place them above the poverty line live in very poor quality housing and usually overcrowded conditions with a great lack of infrastructure and services. At least 600 million urban dwellers in Africa, Asia and Latin America live in 'life and health-threatening' homes and neighbourhoods because of the very poor housing and living conditions and the inadequate provision for safe and sufficient water supplies and for sanitation, drainage, the removal of garbage, and healthcare. If these 600 million urban dwellers are considered 'poor' - for it is largely their lack of income (and assets) that makes them unable to afford better quality housing and basic services - it greatly increases the scale of urban poverty when compared to conventional income based poverty lines.

Although the literature on poverty often refers to people 'living in poverty' it often gives little consideration to their living conditions. The literature on urban poverty pays more attention to housing and basic services than does the literature on rural poverty - and this has been criticised by rural poverty specialists. However, one reason for this is the high cost of housing and basic services in many urban centres,8 which means that housing either takes up a large proportion of the income of low income households or (more commonly) that low income households choose to live in poor quality, insecure accommodation which lacks basic services in order to lower the costs of accommodation. In addition, it may be that government agencies, NGOs and community based organisations are more able to address housing problems and provide basic services to low-income households than they are to change labour markets in ways that increase income-earning opportunities for low-income groups.

International and national figures for the provision of water and sanitation suggest that the inadequate provision in urban areas is not so serious. But official figures exaggerate the proportion of households with safe and sufficient water supplies and with adequate provision for sanitation because the criteria used to define what constitutes an 'adequate' water supply and 'adequate' sanitation are set so low and because governments are given such latitude in defining what they consider 'adequate'. This means that serious questions have to be raised about the validity of tables, so often published in international documents, on 'rural-urban gaps' in the provision of safe water and sanitation. However, it may be that government statistics, published by UN agencies, on the proportion of rural inhabitants with adequate provision for water supply and sanitation are also inflated, for similar reasons.

The Misrepresentation of Urban Poverty

Sent its nature and its underlying causes on how income based poverty lines are often set too low in relation to the income level needed by urban households to avoid poverty They also raised the issue of how well-directed public, private or community action can ensure that those with low incomes can avoid 'living in poverty' by ensuring they can afford adequate quality housing with basic services. This section concentrates on how discussing urban poverty only in terms of inadequate income can misrepresent its nature and its underlying causes.

Most recent critical reviews of official definitions of urban poverty have drawn on the insights of those who worked mainly on rural poverty and who found major gaps between the nature of rural poverty and what official agencies measured. Single poverty lines that divide the population into the 'poor' and the 'non-poor' are often the most inaccurate because they simplify arid standardise what is complex and varied. 'What is measurable and measured then becomes what is real, standardising the diverse and excluding the divergent and different'. The comment by Beck in a book on poverty in village India that '.. if the poor are viewed simply as statistics, figures and ciphers, then the policy that is formulated to alleviate poverty will in all likelihood follow suit and be more relevant to the manipulation of statistics than to the needs of people' also has relevance for urban poverty.

Three points need emphasising with regard to urban poverty The first is that a concept of poverty based only on income fails to pay sufficient attention to the social and health dimensions of poverty (and to other forms of deprivation associated with poverty). The second is the failure to involve 'the poor' in determining what should be done to reduce poverty The third is that equating poverty with income level can obscure the underlying causes and miss the extent to which households face other forms of deprivation due to the strategies they adopt to keep their incomes above the poverty line. Each is considered in a little more detail below.

Poverty and Basic Services

Why are many conventional definitions of who is poor not influenced by whether people have access to health-care, education and adequate quality housing with basic services? Either there is an assumption that these are provided at little or no cost - for instance, through government provided schools and health centres that are open and available free (or at very low cost) to all or at least to low-income groups. Or there is an assumption that those with incomes above the official poverty line can afford to pay for these services and housing. (A third possibility is that people's access to healthcare, education and adequate quality housing is not considered relevant - but this seems unlikely).

Thus, a household with a below poverty line income, whose children go to a free government school, whose family members can get rapid and effective free (or very low cost) treatment when sick or injured, and who live in a shelter with secure tenure and with piped water with adequate provision for sanitation and drainage, is considered to be just as much in poverty as a household with the same income level but who lives in one tiny room or a one-room shack, sharing a water tap and latrine with dozens of other households and with no access to schools and health-care.

One reason for this lack of consideration of access to schools, health-care and adequate quality housing in poverty statistics is the uncritical transfer of 'poverty line' concepts from the North to the South. In the North, when the concept of an income based poverty line came to be applied by governments, a smaller proportion of their (rural and urban) population lived in housing which lacked adequate provision for water supply and sanitation. It is also likely that a much higher proportion of the population in the North had access to free education and to free or cheap health-care at this time. There were thus good reasons why the poverty line income did not include the costs of education and health-care. In addition, the premium that households had to pay for housing with adequate provision for water and sanitation was not great since virtually all the housing stock had such provision. Clearly, households who have to pay for private education and health-care, and a high premium for a house with piped water and adequate provision for sanitation, require a much higher income to avoid deprivation than those who do not have to pay for these.

The scale of the health burden imposed on people who live in very poor quality housing has also probably been underestimated. Inadequate provision of water, sanitation and drainage and poor quality, overcrowded dwellings bring an enormous health burden. Disease or injury burdens per person from diarrhoeal diseases, accidents, tuberculosis, most respiratory infections (including pneumonia, one of the greatest causes of death worldwide) and intestinal worms are so much higher among those living in poor quality, overcrowded housing with inadequate or no infrastructure and services. At any given time, perhaps half of the urban population of the South are suffering from one or more of the main diseases associated with an insufficient or inadequate quality water supply and inadequate provision for sanitation. But very few studies have looked in detail at the health burden faced by those living in very poor quality housing. One that did revealed a tremendous health burden faced by low-income groups in terms of work days lost to illness and injury and the dire economic consequences this brought in terms of increased debt and increased under nutrition for all family members. Yet, most of these health burdens could have been prevented or much reduced at low cost.

The Rights of Poor People to Influence Definitions

The failure of conventional poverty analyses to consider housing conditions and health problems is perhaps part of a more fundamental failure - that of not involving 'poor people' in the definition of poverty and in determining what should be done to reduce it. Low-income urban households may give a high priority to better health-care, better schools, more secure housing tenure, the possibility of home ownership (which for tenants also means avoiding having to pay rent) and improved provision for water and sanitation. Or they may place considerable value on qualitative aspects such as independence, security, self-respect, identity, close and non-exploitative relationships, and legal and political rights, all of which are not necessarily linked to income levels. There are many more aspects of deprivation other than 'income poverty' including vulnerability, powerlessness, isolation and humiliation (ibid.). Without an understanding of the needs and priorities of those who are classified as 'poor' built into the measurement of poverty, the 'poor' will continue to be seen as passive and the 'target' for poverty alleviation.

Poverty Indicators that help Understand Causes

If 'poverty' is equated only with 'lack of income', it can obscure other important aspects of deprivation and their underlying causes. Poverty reduction programmes need to understand these underlying causes if they are to act upon them. The increasing number of detailed case studies on urban poverty has helped increase our understanding of the complexity of the processes that underlie impoverishment and the extent to which these are rooted in the specific social, economic and political structures of these urban areas - although also so often linked to social, economic and political changes at a regional, national and international level.

These case studies point not only to the ways in which households respond to rising prices and/or falling incomes but also to their often negative consequences - the much increased workload on adults (especially women) the pressure on children to drop out of school early so that they can contribute to household income, the increased pressure on households to split up (for instance, one parent going elsewhere where income-earning opportunities are better) and the increasing stress, conflicts and often violence within households. A recent study of household responses to poverty in four poor urban communities not only illustrated these costs but also pointed to other costs related to household responses to rising prices and/or falling incomes. One was the erosion of support networks as the economic crisis made it impossible for some households to fulfill reciprocal obligations. Por instance, when households were coping, they could borrow from each other on a short-term basis - but, with the economic crisis, they felt less able to do so as they were uncertain as to whether they could repay or reciprocate. Also, adult members working longer hours to earn an income meant less time to spend with and supervise children. Often, diets were changed to less nutritious but cheaper foods. And the increased time that women had to devote to income-earning could also mean less time to devote to community management of important services such as health-care or water supply. Thus, many households may succeed in keeping their household income 'above the poverty line' but at a considerable social cost - or at the cost of depleting assets.

Understanding the Need to Act on Multiple Deprivations

If we accept that urban poverty includes not only 'a lack of income' but also other aspects of deprivation such as a lack of assets to help low-income households cope with shocks or stresses, a lack of legal rights (or their upholding), a lack of the resources or contacts necessary to secure political advantage, a lack of access to education, health-care and emergency services and a lack of safe, secure, and adequately sized housing with basic services, then it also greatly widens the potential scope for 'poverty reduction' programmes. But it would be dangerous to downplay the importance of increasing the incomes of low-income urban households. In many urban contexts, increasing incomes for low-income households may be the most effective means of addressing other forms of deprivation by allowing them to increase their asset base, to find (or build) better quality and more secure accommodation, and to avoid having to rely on the more dangerous and illegal occupations and housing sites. The more commercialised the access to those goods and services which allow someone to avoid poverty becomes, the more important access to income becomes.

In addition, many expensive public housing programmes or housing finance programmes implemented during the last 20-30 years, which were justified by the fact that they were 'to benefit the poor' ended up primarily benefiting middle and upper-income groups. The same is probably true of much public expenditure on infrastructure and services - because most new or improved roads, piped water systems, sewers and drains benefited middle income households (or at least not the poorest households). However, there are also many examples of projects and programmes providing or improving infrastructure and services to low income areas of cities that were cheap and effective - with some also recovering much of their costs. There are also many exam- pies of housing finance programmes that allowed low-income households to greatly improve the quality of their homes - at relatively low cost and often with good cost recovery. Improved housing can also imply an important asset, both as a place where income-earning opportunities can be developed and (for the owner) as collateral for credit (ibid.). In effect, there are a range of actions for reducing poverty in urban areas that centre on one of three things:

- Increasing the incomes or assets of low-income households.
- Upholding human rights, including the rights of those with low incomes to be able to call on governments for resources and services and to be able to set priorities both in what is done and how it is done.
- Improving housing and basic services.

Table gives more details. The items marked with an asterisk can be used to strengthen community organisation among low-income groups and to directly or indirectly help strengthen political representation. However, any initiative to reduce poverty that works with low-income groups and the organisations they form in participatory ways can strengthen community organisation and cohesion. This, in turn, can strengthen low-income groups' capacity to negotiate for support or services (or, for those in illegal settlements, for legal tenure) from local authorities and other external agencies.

Acting on Multiple Deprivation

A final point in this book what kinds of response will best address the lack of income and assets, lack of legal protection, the poor housing conditions and other deprivations facing most low-income urban households - whether from a local government agency, NGO, agency of state or national government or international agency.

Table: Different Aspects of Poverty Reduction.

Increasing Income and/or Assets	
A job through employment creation	Where successful, these bring new jobs and/or enhanced incomes, employment creation although external support must understand local constraints on new enterprises being able to generate adequate incomes. Within a stagnant economy with a considerable proportion of the workforce working in activities which generate very little income, there may be little possibility for such enterprises. There may be considerable potential for linking employment creation for low-income groups with public works to improve water supply, provision for sanitation and drainage, improved roads and all-weather paths, health-care centres etc or with staffing new or improved services (eg schools, day care centres, health promoters).
Credit for small-scale or informal enterprise	Examples of credit and support for informal enterprises include the work of the Carvajal Foundation in Cali (Colombia) and of Praja Sahayaka Sewaya in Sri Lanka. There may also be considerable potential for linking income generation and better collection and management of garbage in various cities. Credit for smallscale enterprises must respond to women's needs and priorities, as well as men's.
Education, literacy and vocational training	In general, these should increase income-earning capacity as well as providing other advantages. In many countries, biases against women in education and vocational training will need to be addressed. The barriers to education for low-income households caused by the introduction of school fees or their increase or the increase in other education costs (for instance of school uniforms or examination fees) have to be addressed.
Providing squatters with legal tenure	Increased security of tenure for owner-occupiers' in illegal settlements reduces the risk of eviction, increases the value of their asset and increases the possibility of obtaining credit.
Emergency credit	The ready availability of emergency credit can greatly reduce the vulnerability of low-income groups to economic shocks.
Upholding Human Rights	
Access to justice within the judicial system	This includes legal systems that protect citizens from forced eviction; several million urban dwellers are forcibly evicted from their homes each year, most without any form of compensation. This also includes public programmes to reduce crime and violence within low-income settlements and community programmes to halt the abuse of women and children within families. It is also important in many urban centres to establish the right of low-income urban dwellers to land for cultivation and the need to halt the harassment of hawkers.

The right to vote, to have representative government and to organise to make demands	An increasing number of Southern NGOs who work closely with organisations or federations of low-income groups (People's Dialogue, SPARC, Asian Coalition for Housing Rights etc) have shown the importance of lowincome urban dwellers being able to organise, make demands on the political process and negotiate successfully with government agencies or politicians. Achieving a recognition by government agencies of the civil and political rights of low-income urban dwellers and their entitlements to public support, public services and public accountability can bring major benefits, except where government agencies are themselves too weak and ineffective to do so.
Improving Housing and Basic Services	
Tenure of housing	As well as the advantage noted above in terms of the value of the house and protection from eviction, secure tenure generally promotes household investment in improving the house and gives greater capacity to negotiate with local authorities for improved services.
Improved water, sanitation, drainage and garbage collection"	If adequately provided, this removes a tremendous health burden and also considerably reduces the time needed for domestic tasks. This brings particular advantages to the person in the household who is responsible for collecting water and managing household wastes - usually the woman. Lt is also important in reducing the vulnerability of many low-income settlements to floods and rain induced landslides.
Basic health-care	If available, this greatly reduces the economic and health costs of illness and injury. There are particular advantages for the person in the household who takes care of those who are sick or injured (usually the woman).
Day-care	This increases the time for other tasks for those who look after young children and also means young children are not left in the care of older siblings. Day-care centres can also provide regular health checks for infants and young children and monitor their nutritional status; they can also provide stimulus and support for children's physical and mental development. Day-care centres are often particularly valuable in increasing women's income-earning capacities and especially valuable to single parent (usually women headed) households.
Housing finance	Housing credit available to low-income households who want to build, extend or buy their own homes allows them to afford better quality housing and, if building it themselves, to reduce the time taken to complete it. Credit can also be used to allow improved infrastructure and services for whole settlements - for example, piped water and sewers installed with each household able to repay the capital costs over several years.
Transport	Cheap and efficient public transport can greatly reduce the disadvantages for low-income households of living in peripheral locations and, if citywide, could also help reduce the price of housing.

Many of these deprivations are, in theory best addressed at national, state or city level. 1f the effectiveness of an agency operating at one of these levels is increased and this ensures the provision of good quality health-care, education or water and sanitation in low-income settlements at prices the inhabitants can afford, it reduces the poverty of those living in these settlements. Better public transport across the city will benefit most lowincome groups. Similarly actions by city or municipal authorities can reduce the cost of good quality housing and so allow lower-income groups to afford such housing - for instance by reducing the price of legal land sites (for instance by removing unrealistically high standards for legal land sub-divisions and making the process of land sub-division approval or purchase cheaper and quicker) or by reducing the cost of building materials (for instance removing inappropriate building regulations or ensuring more competition among building material producers).

There are also the problems that arise when poverty reduction concentrates on 'settlements' rather than on the individuals and households who are poor. For instance, not all low-income groups live in the settlements that governments or international agencies designate as 'poor settlements' where improvements can be made. It is common for a proportion of low-income groups to live in very overcrowded conditions and with very insecure tenure in rooms within areas which are nbt seen as 'low-income areas'. In addition, not all groups within the 'poor' settlements will necessarily be low-income groups. And not all will necessarily benefit from improvements made in a 'low-income settlement'. For instance, those who are renting a room, a part of a house plot or a house may find their rent increased when a government programme improves water and sanitation, with the main beneficiary of the government programme, the landlord, now able to get a higher rent.

However, there are at least three reasons why poverty reduction actions should include actions within particular urban settlements in most countries in Africa, Asia and Latin America. The first is the incapacity of so many public agencies to provide the infrastructure and services; a municipal authority may be legally obliged to provide schools, piped water and provision for garbage collection to all settlements but, in most urban areas, they lack the necessary resources to provide these too much of the population (UNCHS 1996). In theory it is probably more cost-effective to address these deficiencies at city or municipal level but it may be impossible to do so (with central governments reluctant to allow municipal authorities more power and resources) or at least to do so in the short term. Or, in prosperous cities made up of many separate municipalities, so many low-income households have become concentrated in the poorest and weakest municipalities while the richest, best provided municipal authorities have long been able to prevent any contribution from their fiscal base to city-wide problems or the problems of low income municipalities. Alternative solutions are needed immediately and cannot wait until more fundamental institutional weaknesses are addressed.

The second reason is the evidence from many case studies that community initiated, directed and managed initiatives are often relatively cheap and may indeed allow significant improvements with most or all costs recovered. The third is the fact that successful community initiated, directed and managed initiatives can empower the inhabitants in the real sense of the word both in terms of their capacity and confidence to address other problems and in their capacity to negotiate with external agencies for resources and support. There is much less hard evidence to back up this point than there is for the other two noted above and it may not be correct in all settlements and societies. One reason for this is that there are so few case studies which follow progress in a predominantly low-income settlement and understand what processes generate improvements.' But it is likely that, as one community initiative in a settlement succeeds and brings benefits to a significant proportion of the inhabitants, so the number of people prepared to support such initiatives increases - and so too does the level of the contribution that many people are prepared to make. It is also likely that as the inhabitants of a settlement become more successful at community initiatives, their capacity to negotiate with external agencies as 'a community' increases. lt is also probable that their feeling of social exclusion is reduced, as different aspects of the deprivation associated with social exclusion are lessened or removed. The lives of those living in one particular settlement can be transformed by a process that, over time, allows them individually and collectively to improve their incomes, their housing and living conditions, and their access to health-care and education. This process can also mean increasing social inclusion - as, for instance, the inhabitants negotiate with external agencies to take over the management of the water system, school or health centre they constructed themselves some time earlier, when there was little or no hope of getting public agencies to provide these. In addition, because this is a process largely directed by the inhabitants (although influenced by what is possible and what resources can be obtained), it allows them to set their own priorities as to which aspects of deprivation are addressed first.

Increased income will generally be among the top priorities of low-income households although it may be more difficult to negotiate support for income-generating activities from an external agency, or it may be very difficult to find economic activities that provide higher or more stable incomes. There are also important income-earning benefits from action in other areas. For instance a well-managed day-care centre allows the people in households who look after infants and young children more freedom to earn an income; or a housing loan programme which allows a household to extend the house to include a small business or to afford to link up to the electricity grid to use power tools; or a health programme that greatly reduces the amount of time people are sick or injured and the amount of time that has to be spent by other household members nursing and caring for them. In addition, expanding or improving provision for day-care, schools and health centres generally means increased employment - not only of the professionals but also support staff - and the funding to cover these salaries may be more easily obtained from local sources than funding for employment creation.

The most appropriate mix of these actions will vary greatly, according to the circumstances within each settlement and to the priorities of its inhabitants. This underscores the importance of involving the inhabitants of each settlement in determining priorities. And, of course, priorities will differ among different groups within each settlement. This underscores the importance of ensuring that all inhabitants' needs and priorities are considered. The fact that it is usually women within low-income households who take on extra work and responsibilities when incomes fall or prices rise also means that their needs and priorities should receive special attention within poverty reduction measures.

If we consider the process that removes the many aspects of deprivation suffered by those living in one particular settlement, there are three important characteristics to consider. The first is many needs (e.g. in one particular settlements perhaps new job possibilities, a health centre, day-care provision, improved water and sanitation, repairs to the local school and more textbooks and basic equipment, a more secure and certain legal tenure of the land site, cheaper building materials, etc.) with different groups within the settlement having different priorities. The second is different possibilities for getting action on these needs - for instance there may be a good possibility of getting funding for a health centre but very little of getting piped water. The third is (generally) a range of external agencies with whom the inhabitants can negotiate for funds, services or technical or legal advice - and each external agency has their own possibilities, priorities and limitations. Most government agencies have responsibility for only one sectoral intervention; there is no point in negotiating with the department of education for a health centre. Most international agencies also have their own pre-defined priorities; there is no point in seeking support for a day-care centre if the agency concerned does not consider this a priority. There is also the division between capital costs and running costs; some agencies will only cover one or the other.

So, what kind of external agency could support this process of community directed and managed action to reduce poverty in settlements? From what has been stated above, it would need to be able to:

- Support a diverse range of initiatives in different sectors in a large number of predominantly low income settlements and have the capacity to provide very small grants and loans.

- Allow considerable autonomy to the inhabitants of each settlement in setting priorities, in determining how these are addressed and in implementing or managing actions.
- Recognise the need to support and reinforce long-term processes within each settlement so that the fact that the inhabitants of one settlement receive support for one initiative does not disqualify them from receiving funding for another.
- Encourage solutions that keep costs to a minimum.
- Encourage cost recovery wherever possible (and obviously, if solutions keep costs to a minimum, this also makes cost recovery more feasible).
- Respond rapidly to requests for funding.
- Provide or fund technical and legal assistance where needed to allow all the above. If this is combined with concentrating on keeping capital costs low, on supporting a great range of projects (including very cheap ones) and ensuring that the settlements with the least organised community organisations are helped to act, the external funding agency will probably have to accept that a relatively high proportion of its project expenditure will be taken up by staff costs.

Such an agency should also, ideally, be transparent in the sense that all groups in a city who apply for funding or technical support should be able to know who else applied for funding, who gets funding and why - and who does not and why.

Unfortunately, these characteristics, which seem appropriate for supporting settlement level initiatives for poverty reduction in urban areas, are the opposite of the characteristics of most official donor agencies and multilateral banks who seek to support poverty reduction in urban areas. Most such agencies were set up to fund relatively large capital projects on the assumption that there are local government institutions able to implement them. Most such agencies assume that they will not get the funding back, which discourages the search for cheap solutions and solutions where costs are recovered. Most are reluctant to fund recurrent costs. Most such agencies are also under great pressure to reduce staff costs and to keep to a minimum the proportion of their total expenditure that goes on staff costs. The idea that an 'efficient' donor agency is one that spends a lot of money with a minimum of staff costs remains a widely held one among donor governments and, in general, by the citizens to which they are accountable. This suggests that making international agencies more successful in supporting community-directed actions for poverty reduction in urban areas implies not only new priorities but new modes of operation. Alternatively, if the donor agencies' institutional context does not allow them to change, they will need to make greater use of intermediary institutions located within recipient countries which can have the characteristics that are noted above. For instance, a large number and range of community level initiatives could be supported within each city by a Fund for Community Initiatives based in that city, funded by international donor agencies.

References

- CITY-PLANNING-IN-GRAECO-ROMAN-TIMES-WITH-EMPHASIS-ON-HEALTH-FACILITIES-280210312: researchgate.net, Retrieved 25, August 2020
- MESOPOTAMIA-THE-CRADLE-OF-CIVILIZATION-AND-MEDICINE-333533736: researchgate.net, Retrieved 18, February 2020

2

Urban Growth: An Integrated Study

The rate at which the population of an urban area increases is termed as urban growth. Some of the major issues connected with urban growth are overpopulation, unemployment, sanitation problems and shortage of water. This chapter discusses in detail these problems related to urban growth as well as the stages of urban growth.

Urban growth refers to the rate at which the population, land area, or significant land-use increases. Also defined as the growth of metropolitan areas or cities, it is something that started way back between the 5000 and 6000 B.C. especially due to the movement of people from rural areas to urban areas – places with high human population, economic activities, and infrastructure. The 19th century, in particular, was a mark of actual urban growth.

And since 1800, the urban revolution has rapidly increased, reaching new levels in the history of the world today. Urban growth is also closely linked to urbanization or urbanism, a term used to refer to an increasing proportion of a population residing in urban areas such as cities, suburbs, towns, and contributions.

Urban growth is, therefore, utilized as an indicator for a country's or an area's economic condition as well as development. The growth of urban areas is often influenced by certain factors such as surplus resources, development of infrastructure, commercialization, education, and mining, among others.

Various Causes of Urban Growth

The Natural Increase in Population

The rate of death and births characterizes the natural expansion of an area. In areas where births are more than the deaths, the population is bound to increase. People who migrate to town and cities tend to be young people who are in search of housing, jobs, or better education. Young men and women have a high fertility rate; therefore, they increase in numbers quickly and will eventually look for new spaces within the urban area to settle and fend for their kind.

Migration

Immigration is a major contributor to the increase in the population of a place. Many people are forced to move to urban areas in search of jobs, education, and housing. Inadequate funding and social infrastructure have also driven people to urban areas. In other cases, political, racial, economic, or religious conflicts have forced people to move to neighbour urban areas.

Industrialization

The industrial revolution brought about new production techniques. By this, manufacturing has created more job opportunities by providing people with the opportunity to be employed in new sectors. With modern farm machinery, employment in the rural areas reduced, forcing workers to move in search of new jobs in tertiary and manufacturing industries.

Also, in pursuit of increased wages, men and women continue to abandon jobs in rural areas by moving to industrial cities. Many people move to urban centers to look for well-paying industrial jobs as urban centers have uncountable opportunities.

Commercialization

Whereas industrialization has played an essential role in the growth of urban areas, commerce and trade have profoundly influenced the growth of cities. In ancient times, cities like Athens, Sparta, and Venice were great commercial centers.

In modern life today, the commercial activities in city and industrial areas continue to attract more and more people as traders and workers, thereby contributing to the growth of cities and town areas.

Businesspeople prefer going to the cities to sell their commodities and in search of higher profits. The kind of economic pull urban centers have is one that attracts more and more people to move to urban areas.

Advancement of Transport and Communication

When factories were introduced, the local transportation was weak, forcing the factory laborers to reside near their place of work. The cities were partitioned into dwelling areas, market area, factory area, slums, and so on. Increased population led to congestion of housing and added to the already existing community by extension of boundaries.

Today, people prefer to live near their place work not because of poor transport but to shorten distance travel and avoid the traffic jam on their way work. Active transportation helps make the cities more habitable by easing communication, transportation, and creating convenient accessibility.

Availability of Educational and Recreational Facilities

Most training institutions, colleges, elementary schools, and technical institutions are in urban areas. Also, most libraries are located in the cities. Recruiting agencies, as well as the examination councils, are situated in the cities.

Clearly, due to the location of these facilities, most students and adult learners are attracted to the town for easy accessibility of higher education. Opera and Amusement Theaters are also in urban areas, drawing more people to the cities.

Urban Planning Policies

Urban planning typically forces city leaders to find ways of making a sustainable city, including proper town development and expansion plans. As a strategy aimed at converting vision into implementation, it acts as a guide for making the most out of a city by improving infrastructure and building its economic growth and enhancing the city's living standards as well as the resident's well-being. With better living standards and infrastructures in the city, more people are attracted to live in the municipality, consequently increasing its population.

Topographical Factors

The topography of an area can have huge impacts on the growth of a city. An area with a suitable topography is usually easy to develop and expand. Urban areas in or around an excellent topographical area are easily extended and refined, thus drawing more people to such areas.

Unbalanced Spatial Development

In many countries around the world, there are developed regions and less developed regions. Many people tend to move to more developed areas. Occasionally, urban centers (cities) are more developed than rural areas because of the available facilities and opportunities offered in urban areas. On this account, the population and size of the metropolitan area will automatically increase within a given period.

Transformation and Modernization in the Way of Living

Transformation and modernization play a very significant role in attracting people to the cities. As technology improves, together with highly sophisticated infrastructure, liberation, communication, dressing code, medical facilities, and other social services offered; people tend to believe they can live a more comfortable life in the cities.

In growing urban areas, people easily embrace changes in the styles of living, for instance, mode of dressing, attitude, habits, and views. As a result, more people are attracted to the cities, and the cities continue to grow day after day.

Mining and Investments

As more urban areas continue being successful and well-off due to the discovery of minerals, resource exploitation, agricultural activities, or business operations, urban areas continue to grow economically. Societies continue being financially liberal due to the availability of jobs in urban areas. As a result, more investors are attracted to invest in such cities, promoting their growth.

Presence of Functional Administrative and Social Support Offices

The presence of administration dominance in an urban center promotes its growth. Many cities today hold offices for various administration and governance functions, which influence people to live in these areas for access to better governance and social amenities like security, sanitation, housing, education, and healthcare.

Differences between Urban Growth and Urbanization

The difference between urban growth and urbanization can be categorized using three broad aspects.

Population (Immigration Versus Rural-urban Migration)

Urban growth is characterized by population increase, while urbanization can be termed as the increasing proportion of a country's population living in an urban area. Urbanization is expressed as a ratio or a percentage of the entire population, through the process of transforming rural areas into urban areas.

Urbanization, therefore, mainly involves rural-urban migration. Contrarily, urban growth involves the immigration of people. Urban growth happens when the increase in the number of people is not proportional to the number of urban dwellers in an area.

Land Area (Extensive Land Area Versus Intensive Land Area Coverage)

In terms of urbanization, the land area covered is often extensive, owing to large-scale economic development. In other words, urbanization is linked to economic development in the sense that urbanization cannot occur if there lacks widespread economic development that covers a wide geographical area, which often comes about due to high per capita income. Urban growth, on the other hand, can take place without far-flung economic growth and it is not strictly influenced by extensive land area coverage.

Infrastructural Versus Economic/Commercial Change

Urban growth primarily happens when there is infrastructural development in terms of manufacturing, invention, housing, and new transport systems or routes that encourage the immigration of people into new urban spaces.

Urbanization, on the other hand, is mainly influenced by increased commercial or economic activities within an area. The meaning of this is that the economic hub of an area defines urbanization. The more the economic, capital investments, and trade activities within an area, the more urbanized it becomes.

Effects of Urban Growth

Increased Air Pollution

Urban sprawl increases car, bus and truck traffic by creating longer and more frequent commutes, which leads to a major increase in air pollution and ground-level smog. Vehicles using fossil fuels are the number one cause of air pollution in many urban areas with serious implications for public, wildlife and ecosystem health.

As more time is spent in cars and more traffic congestion occurs over a larger area, it contributes to the growing emissions of greenhouse gases and particulates, resulting in the continued degradation of air quality in urban areas.

Increased Water Pollution

Urban sprawl is the cause of water pollution as rainwater picks up gasoline, lawn chemicals, heavy metals, paints spills, motor oil, pet wastes, construction site erosion and other pollutants in runoff from lawns, driveways, roads and parking lots, which can eventually travel in large, concentrated amounts, polluting nearby water sources, such as a stream, river or lake.

High Water Consumption

Urban sprawl can create water distribution issues and lead to water over-consumption as more water is consumed for lawn watering and other landscape activities, which can strain and deplete local water supply systems. According to the EPA, "an American family of four can use 400 gallons of water per day, and about 30 percent of that is devoted to outdoor uses. More than half of that outdoor water is used for watering lawns and gardens.

Nationwide, landscape irrigation is estimated to account for almost one-third of all residential water use, totaling more than 7 billion gallons per day. Other residential outdoor uses include washing automobiles, maintaining swimming pools, and cleaning sidewalks and driveways."

Health Degradation

As communities are auto-dependent, we need to drive to schools, shops, parks, entertainment, playdates, etc. Thus people become more sedentary, making it more difficult for them to get physical activity and maintain a healthy weight. Residents of sprawling counties were likely to walk less during leisure time and weigh more than residents of compact counties.

A sedentary lifestyle increases the risk of overall mortality, cardiovascular disease, and some types of cancer. The effect of low physical fitness is comparable to that of hypertension, high cholesterol and diabetes.

Loss of Open Space and Wildlife Habitats

Urban sprawl threatens productive farmland, transforms parks and open spaces into highways and strip malls and destroys more than one million acres of parks, farms and open space each year. The increased proximity and accessibility of urban activities to natural areas impose stress on ecosystems and species through the noise and air pollution. The natural habitats of wildlife are disappearing beneath the concrete, which is threatening important ecosystems around the world.

Sprawl is claiming farmland at the rate of 1.2 million acres (10.5 million hectares) a year. Throw in the forest and other undeveloped lands, and, for a net annual loss of open space, you're waving good-bye to more than two million acres (10.8 million hectares)."

Increased Traffic Congestion

As urban areas spread out, trip times are lengthened, residents are forced to virtually drive everywhere like drive to schools, work, shops, parks, entertainment, play dates, etc., spending more time in their cars and trucks, and traffic congestion occurs over a larger area.

Increases Risk and Damage from Floods

Sprawling developments drain and destroy wetlands that absorb floodwaters, and can be built in floodplains, leads to a higher susceptibility to floodwaters. In the last eight years, floods in the United States killed more than 850 people and caused more than $89 billion in property damage. Much of this flooding occurred in places where weak zoning laws allowed developers to drain wetlands and build in floodplains."

Stages of Urban Growth

It may appear that urban centres have experienced a simplistic growth from being the smallest to the largest though an inherent unidirectional process. But scholars over the time have come up with different theories to track the growth of the urban nodes through sequential stages. Theoretically several models have tried to explain the stages of urban growth. The classic structural change model identified characteristic features of the development process, such as the shift from agricultural (primary) to industrial (secondary) production and characteristic phases of urbanization on the basis of logistic patterns in urban growth. On the other hand, some models have even tried to phase out urban growth on the basis of GDP per capita, industrial structure (the proportion of the economy devoted to primary, secondary, or tertiary industry), employment structure (the proportion of jobs associated with each industrial sector) and level of urbanization (the proportion of non-farm population in cities) as indicators of the stage of city development. However, different urban geographers conceptualised the urban growth in well sequenced stages and generated a model of urban growth on the basis of population growth in the urban regions and the shifts in population between the core and the peripheral zones. Population mostly focuses on the core for their economic activities. Berg's model proposed that the urban regions develop in well deciphered stages of Urbanisation, Suburbanisation/Ex-urbanisation, Counter-urbanisation and Re-urbanisation.

Theories of interdependency, mercantilism and that of global urban development have sought to propose a kind of causal between processes of wealth accumulation and the creation of hierarchy of the urban spaces. These have further talked of urban development as consequent upon the stages of economic development from mercantilism through industrial and monopoly capitalism to corporate capitalism. In the later stages of economic development the urban spaces at the macro level are open to exchange of both human and social capital. However, if the same is thought at a much smaller scale i.e. at a meso level, then such hierarchies are also found to exist within urban regions of particular nation.

Researchers have employed the concept of cyclic urbanization wthin individual urban agglomerations and have identified four major stages in which they grow. The stages may be summed up as:

- Urbanisation,
- Suburbanization or ex-urbanisation,
- Dis-urbanisation or counter-urbanisation,
- Re-urbanisation.

While in the stage of urbanization, certain settlement grow and expand at the cost of their surrounding countryside, the sub-urbansiation or the ex-urbanisation phase experiences the growth of the peripheral ones of urban settlements at a higher rate in comparison to the core. In the stage of counter urbanization, the core experiences a loss of urban population which far exceeds the rate of population gain at the peripheries resulting in the overall loss of the urban settlement in terms of population numbers and in the stage of re-urbanisation, the core starts gaining population whereas the ring keeps on loosing population.

STAGES OF URBAN DEVELOPEMENT

		Population change characteristics			
Stage of development	Classification	Core	Ring	DUS	
I.Urbanization	1 Absolute centralisation	++	-	+	Total growth (Concentration)
	2 Relative centralisation	++	+	+++	
II.Exurbanisation	3 Relative centralisation	+	++	+++	
	4 Absolute centralisation	-	++	+	
III.Counterurbanization	5 Absolute centralisation	--	+	-	Total growth (Deconcentration)
	6 Relative centralisation	--	-	---	
IV.Reurbanisation	7 Relative centralisation	-	--	---	
	8 Absolute centralisation	+	--	-	

Stages of Urban Development after van den Breg.

However, Breg's model is based on the changes in the rate and direction of population movement between the urban core and the peripheral ring which together form the daily urban system (DUS) which together are functionally interrelated. Absolute change is noted in the urban settlement in cases the core and the ring experience change in population in two different directions and relative change is experienced when the population moves in both the zones but at different rates but in the same direction. The stages with their relative change in the direction of population have been summed up in the following figure.

Geographers postulated the phases of urban growth in their theory of Generalised Stages of Differential Urbanisation. They identified that the cities grow through the following stages of:

- The primate city phase.
- The intermediate city phase.
- The small city phase.

In the initial primate city phase, population growth and economic activities are concentrated in the rapidly growing primate cities. The cities slowly attain their multinodal character through the stages of early primate city, intermediate primate city phase and advanced primate city phase.

On the other hand the intermediate city phase is characterized by the slowing down of the growth rate of the primate city and spatial de-concentration. This phase experiences the growth intermediate sized cities in close proximity to the primate city. The small city phase on the other hand sees the growth of small urban centres which grow at a much faster rate in comparison to the primate city and the intermediate cities.

Urbanization and Economic Growth

Recognition that urbanization has a role to play in economic and social progress is important, bearing in mind the general neglect of cities in the national and international discourse on development in recent decades. However, to avoid the pendulum of fashionable thinking simply swinging back towards rural development in a few years' time, it is vital that the supporting rationale and evidence base are robust. The current arguments put forward by these development organizations for urbanization being a driver of development, or for cities functioning as "engines of growth", are broad-brush and the supporting evidence provided is rather thin. The urbanization–growth nexus is often portrayed as automatic and inevitable, like some sort of universal law governing an immutable historical process. Insufficient attention is paid to the dynamics of urbanization and the nature of growth, including the different ways in which cities can support growth and the varied forms or composition of that growth. The conditions necessary for this to occur are also neglected, along with the consequences of growth for poor communities and the depletion of natural resources. Urbanization is often conflated with agglomeration, but they are not synonymous, especially if the fastest-growing areas are small cities and towns rather than major cities.

China is widely held up as an example of how urbanization can fuel industrialization and transform living standards. China passed the historic milestone of 50 per cent of its population living in cities in 2011, up from only 20 per cent in 1980. The unprecedented speed of urbanization has reflected the strength of jobs growth in cities. Average household incomes in Chinese cities are now almost three times higher than in rural areas, largely because of higher productivity. The state's commitment to invest heavily in urban infrastructure has helped to limit the socially disruptive effects of this massive population movement. It differs from India and many other developing countries, where urban congestion, water shortages, squalid living conditions and public health problems are rife. In the early 2000s, China spent 12.6 per cent of GDP on infrastructure – more than twice that of India. "For every pound Indian authorities invest in urban infrastructure, their Chinese counterparts spend seven."

China's experience seems to bear out the broader argument developed in this paper that it is not urbanization per se that stimulates and sustains growth, but rather the form that urbanization takes and whether it provides an efficient enabling environment. This includes the manner in which the growing population and economic activity are accommodated within cities and the distribution of this growth across cities of different sizes and across the wider national territory. Public investment in infrastructure and the built environment can play a crucial role in shaping these patterns, and therefore in influencing whether or not urbanization is functional for development. This will also influence the impact of urbanization on poverty and the environment.

The Theory of Agglomeration Economics

Two concepts underpin our understanding of the advantages of economic concentration – the division of labour and economies of scale. The former explains the gains for productivity, and therefore growth, from specialization. Firms organize themselves around particular products or tasks, which yields efficiencies and enhanced skills. Specialization also applies at the city level, meaning there are benefits of focusing on a function, or group of functions, for which places inherit or build a distinct advantage. Specialization becomes more important as external trade grows and competition intensifies.

Economies of scale have two aspects. Internal economies of scale are internal to the firm and relate to the lower unit costs or efficiencies that result from larger-scale production. External economies of scale (or "agglomeration economies") are the benefits firms obtain from being close to other firms in order to reduce transaction costs (such as transport and communication) and to gain from network effects, such as shared information. The bigger the network, the more knowledge and intelligence are available to learn from. Agglomeration economies include proximity to a large labour pool, suppliers, customers and competitors within the same industry (localization economies), and firms in other industries (urbanization economies).

These advantages can be summed up as three broad functions: matching, sharing and learning. First, cities help firms to match their unique requirements for labour, material inputs and premises better than towns, because markets are larger and there is more choice available. A bigger pool of providers also lowers costs and improves variety through competition and specialization. In volatile market conditions there is a premium on being adaptable, and companies tend to be leaner, more focused on core competences, and more reliant on buying-in goods and services than in-house production. Agglomerations assist firms to "mix and match" their resources more easily.

Second, cities afford firms access to a wider range of shared services and infrastructure because of the scale of activity. Cities offer better external connectivity to customers and suppliers through more frequent transport links to more destinations and more efficient logistics systems to handle imports and exports. They tend to have higher-capacity telecommunications systems for electronic connectivity and marketing. There is more extensive professional expertise, financial know-how and specialized research available to help companies stay up-to-date with changes in technology.

Third, firms gain from the superior flow of information in cities, which promotes more learning and innovation, and results in more valuable products and processes. Proximity allows face-to-face communication and sharing of complex ideas between companies and support organizations. It enables people and firms to compare, compete and collaborate, which can establish a self-reinforcing virtuous circle that spurs creativity, attracts mobile capital and talent, and generates growth from within. These "dynamic" advantages become increasingly significant over time because they are cumulative, unlike the one-off or "static" advantages of lower production and transaction costs.

Economies of scale also apply to public infrastructure. It is more efficient to provide hospitals, sewage treatment facilities and universities in large cities than in dispersed settlements, where densities are low and distances are large. Similarly, urban infrastructure projects generate greater positive externalities in cities than in towns because they increase the productive capacity of households (through improved health and life expectancy) and the effective labour supply, which enable higher economic growth. Researchers provide examples of how public investment in low-cost housing, sanitation, drinking water, electricity and access roads in informal settlements in Africa helped to reduce public health problems and mortality associated with overcrowding. They argue that governments must look beyond individual projects when assessing their costs and benefits because of the disproportionate gains associated with urban investments.

The existence of agglomeration economies does not mean that urbanization will necessarily cause economic output to increase. The benefits of concentration can be offset by rising congestion, overcrowding, overloaded infrastructure, pressure on ecosystems (such as water courses and air quality), higher costs of living and higher labour and property costs in cities. These negative externalities tend to increase as cities expand, especially if urban development is haphazard and there is insufficient public investment to maintain and expand essential infrastructure. Dysfunctional systems, gridlock, power cuts and insecure water supplies increase business costs, reduce productivity and deter private investment. The balance between the agglomeration economies and diseconomies may have an important influence on whether city economies continue to grow, stagnate or begin to decline.

The advantages of agglomeration also vary across sectors of the economy, with implications for the relationship between urbanization and growth. Economic success generally involves a progressive shift out of agriculture, even as agricultural productivity grows, and the diversion of resources towards industrial and service sectors. Agriculture is more land intensive and gains from being dispersed across the countryside, while industries and services benefit from concentration. This helps to explain why urbanization is quite closely related to the share of a country's GDP in industry and services. It does not in itself imply that faster urbanization will lead to accelerated economic growth. Rather, it suggests a sort of equilibrium or mutually dependent relationship, wherein urbanization helps to sustain growth but could be counter-productive if it outpaces economic restructuring and makes it difficult to absorb the surplus urban labour force.

Just as important from a policy perspective, the benefits of agglomeration depend on specific investments being made by local and national governments to improve how cities function. Governments have a big influence on which cities receive favourable treatment. There is little point in building new roads if they worsen congestion elsewhere in the city, or opening sewage treatment works or power stations if they discharge effluent and pollution in ways that discourage further investment in the city. The gains of agglomeration also depend on the mobility of enterprises and workers to productive locations to facilitate matching, sharing and learning. The agglomeration literature has tended to assume that land and housing markets function smoothly and local authorities are responsive to market failures, so that people and firms can easily relocate and suitable land and infrastructure will be provided to accommodate them. This is far from straightforward in many developed as well as developing countries, because all sorts of other factors are also at work.

Finally, urbanization implies a decline in the share of rural dwellers, and not just more urban dwellers. This may be caused by a rise in rural productivity, or it may require such a rise. Where rural overpopulation is a problem, urbanization can assist. Urbanization can also reduce the economic distance between rural producers and their markets by improving the connecting infrastructure. Not all forms of urbanization will have the same impacts on rural productivity and incomes. Much will depend on how many cities are expanding, whether they are spatially concentrated or dispersed across different regions, what activities they undertake and what sort of infrastructure they invest in.

Measuring the Relationship between Urbanization and Growth

It is difficult to disentangle and measure the effects of agglomeration because of their complexity and feedback effects. The benefits of economic concentration are partly absorbed by higher land and labour costs and offset by increased congestion. Consequently, the effects may not be readily apparent in aggregate economic indicators such as output, employment or average earnings. These variables are influenced by many other factors as well, such as the city's industrial structure and occupational composition. The effects may not be apparent at the scale of city administrative units, for which spatial data is conventionally available, because of their openness and leakage of resources across boundaries. Government financial transfers between localities can also mask the underlying economic processes, particularly when they are specifically designed to redistribute from stronger to weaker local economies. There are additional complications of timing in assessing the dynamic effects of agglomeration.

From the perspective of economic growth, the key outcome of agglomeration that should be measured is productivity. This is the single most important determinant of growth in economic output and income. It reflects the market value of local goods and services and the efficiency with which they are produced. Unfortunately, statistics on productivity tend to be unreliable at the local level. All sorts of other devices and indicators have been used instead to analyze agglomeration advantages.

One of the simplest is to compare the urbanization levels of individual countries with some measure of national economic output, average income or social development, in order to test for a statistical association between urbanization and development. Many results suggest that there is a broad empirical regularity – highly urbanized countries tend to be more prosperous. Researchers found a strong correlation (R^2 = 0.57) between the proportion urbanized and Gross National

Income per capita in 2004 for countries around the world. They argue that the correlation would be even stronger if there were fewer international differences in the way "urban" is defined. Yet the spread of observations around the trend line is always very wide, indicating that other factors are also involved. In addition, the discovery of a statistical association does not of course constitute evidence of a causal connection. Urbanization may be as much a consequence of economic development as a cause, and there may be different levels of urbanization suitable to different economies.

Some other independent processes may also contribute to both urbanization and higher levels of development in cities. The concentration of political power is an obvious example, resulting in "rent seeking" rather than productive activity, as people and firms seek preferential access to political elites. The outcomes could include more public sector jobs and higher wages in cities, plus higher levels of professional services, construction, embassies, media and consultants seeking access to government contracts and lobbying over legislation. This has been the dominant factor in the growth of many capital cities in Africa: "Proximity to the centre of patronage and redistribution, rather than economic development, has driven the explosive growth of African capitals since independence." "The more centralized a nation's government, the larger its capital city, because people are attracted to power as ants are to picnics."

A series of other studies involving correlations between some measure of agglomeration (such as city size or density) and some measure of economic development (such as average income or growth in output) have found that no relationship exists (e.g. studies summarized in Martin). Leading economic geographers acknowledge that the present understanding of the dynamics of agglomeration and the strength of the cause effect mechanisms remains quite limited, despite considerable theoretical development. For example:

> "All in all, work on agglomeration has progressed considerably in the NEG [New Economic Geography], urban economics and regional science. But it remains far from an adequate causal account of the dynamics of agglomeration and de-agglomeration Deciphering the causes of agglomeration will ultimately require a much better understanding of the complex interdependencies between agents that lead them to congregate together."

Serious weaknesses in the main theories of agglomeration arise because they embody crude conceptions of geography and history." One of the consequences is that they are unable to explain differences in the strength and nature of agglomeration effects in different places and at different points in time. Despite being one of the leading proponents of formal economic models based on many simplifying assumptions, "Not only is there no one formula toward urban eminence, but also the sources of success are often highly nation specific."

With the focus of economic research on the relationship between cities/urbanization and productivity/growth, there has been little systematic analysis of the varied forms or composition of urban economic growth. In particular, the problems of narrow, unequal and exclusive urban growth have not been examined closely. This gap is a particular concern in Africa, given the recent revival of many national economies based on booming commodity exports. The imperative to stimulate growth of any kind has been replaced to some extent by the need to diversify and broaden the growth trajectory in order to increase the impact on jobs, incomes and poverty. Cities could be a key part of this agenda by providing the enabling environments for industrialization through essential infrastructure, skills, technical support, investment and responsive governance.

Another area of neglect has been the implications of urbanization for ecological systems and the use of natural resources. In the context of climate change, volatile food and energy prices, and global pressures to cut carbon emissions, it is vital to improve understanding of the environmental consequences of fast-growing cities. At first sight, these concerns should reinforce the case for concentrated rather than dispersed development, including localized production of energy, food, water, building materials and many other goods and resources. This will reduce transportation, economize on the use of land and improve the efficiency of resource use. As with economic growth, however, there is some debate about the environmental costs and benefits of urbanization. Much is bound to depend on the form of urban development, including the issues of compactness, mixed land uses, public transport, and design for energy saving and recycling waste, water and other resources.

Evidence of Agglomeration Economies from the Global North

A range of more sophisticated econometric studies in the United States and Europe have been carried out in recent years in order to estimate the effects of agglomeration. The emphasis has been on cities of different sizes rather than urbanization per se. They use different statistical variables, units of analysis and measurement techniques, and their findings are unsurprisingly diverse. Many of them conclude that cities do offer measurable economic advantages, although they are not as substantial or widespread as often suggested by agglomeration theory or by policy proponents.

A useful summary of the evidence concluded that the elasticity of city productivity with respect to city size is somewhere in the range 0.03–0.08. This means that doubling city size increases productivity by 3–8 per cent. Put differently, for an increase of 25 per cent in a city's population, the output per worker (and consequently income) rises by 1–2 per cent. Studies based on individual earnings data find somewhat smaller, although still significant, impacts of agglomeration in big cities. Several other studies suggest that the impacts are larger for cities that specialize in particular industries, indicating that localization economies may be stronger than urbanization economies. They found that the elasticity of productivity ranged widely between 0.01 and 0.20, although most were under 0.10. This is modest rather than substantial. The impact of other drivers of national economic growth, such as trade, technical change and investment in physical and human capital, are likely to be more important.

Table: Estimates of agglomeration economies in the North.

Estimates of Agglomeration economies in the North		
Unit of analysis	independent	Elasticity
Swedish cities	City size (population)	0.02
US MSAS	City size (population)	0.20
US MSAS	City size (population)	0.06
US MSAS	City size (population)	0.20
US MSAS	City size (population)	0.10
US MSAS	City size (population)	0.03
US MSAS	City size (population)	0.05
US MSAS	City size (population)	0.07
Japanese cities	City size (population)	0.03

Japanese cities	City size (population)	0.04
Greek regions	City size (population)	0.05
US MSAs	City size (population)	0.01
Japanese cities	Industry size (employment)	0.05
Brazilian cities	Industry size (employment)	0.11
US MSAs	Industry size (employment)	0.19
US MSAs	Industry size (number of plants)	0.03
US states	Employment density	0.06
EU regions	Employment density	0.05
Britain sub-regions	Economically active population	0.05
British cities	Employment density	0.02

Researchers also measured the rate at which the gains from proximity diminish with distance from the core city. They found that the advantages are greatest within 40 minutes driving time of the central city, tapering off quite sharply thereafter and having little or no effect beyond about 80 minutes. The agglomeration effects are four times stronger 30 minutes away than 60 minutes away. This finding has important implications for urban form and investment in infrastructure. It suggests that dispersed patterns of urban development undermine productivity and growth by lengthening commuting times and opportunities for face-to-face interactions. It also justifies transport improvements to reduce travel times and traffic congestion in the largest cities. This was the main reason why a major UK government report concluded that the top priority for national transport investment was Greater London. It challenged the fashionable idea of connecting different cities across the country through high speed rail in favour of making mobility across London more efficient.

Another recent UK study disaggregated the effects of agglomeration for different industries and found somewhat larger effects, with the elasticity of productivity varying between 0.07 for manufacturing and 0.19 for services. The effects were small or insignificant for non-tradeable services such as retailing, real estate, postal services and public services, which tend to locate close to consumers. The biggest effects of all were for financial services and professional services such as management consultants. This chimes with recent qualitative research across eight European countries on high order business services (lawyers, accountants, marketing, insurance companies, engineering consultants and designers). They found strong tendencies for these top of the food chain functions to concentrate their activities within each country's largest city.

Summing up, econometric studies have yielded diverse findings for the magnitude of agglomeration economies. Some of this is bound to be attributable to methodological differences and varied national circumstances. Whatever the reasons, the weight of evidence is not substantial enough to justify popular assertions that cities are "engines of growth". "We simply do not know enough to justify yet further concentration of economic activity in already congested and overheated regions and agglomerations." Another recent European study concurs that:

> "Many of the standard textbook arguments no longer hold in many parts of Europe large cities no longer play the driving role a policy just focusing on large urban areas in the EU will likely miss its target in terms of maximizing aggregate growth size is not a good predictor of growth."

A recent UK study also raised doubts about the special advantages of cities: "There is no overwhelming evidence in support of British cities being the ideal locations for encouraging growth."

One concern is the level of aggregation. Most previous studies are based on city level data, which may be too highly aggregated to reveal the way agglomeration economies operate. These are essentially about the interactions between individual firms, workers and supporting institutions. More disaggregated analysis based on data at the level of industries, firms and small areas might reflect the underlying processes more clearly. They might also identify the industrial and geographical scope of agglomeration economies more accurately, including the significance of the spatial structure of cities and role of efficient infrastructure. Agglomeration effects should also not be examined in isolation of other factors affecting productivity and development, such as the industrial and occupational structure, level of technology and skills, available land supply, and internal and external connectivity. Some of these forces are likely to reinforce each other, with successful cities emerging as the sum of the parts.

Urban Growth Problems

Currently, half of the total population of the world lives in urban cities, a trend that is definitely going to continue for years to come. Despite urbanization creating opportunities for people who take the leap to look for greener pastures in urban areas, it is often faced with a lot of challenges which this article looks at in-depth.

Overcrowding or Overpopulation

Overcrowding is a situation where a lot of people accumulate in a rather limited space that is unable to accommodate them without succumbing to the pressures around it properly. As a result of urbanization, overcrowding is a persistent problem as a large number of people are consistently moving to urban areas on a daily basis.

This leads to cities growing in population and getting crammed when it gets beyond its capacity. When a city is at maximum or excess capacity, the people tend to compete over the limited and scarce resources such as electricity, water, transport, and the main reason they make a move, employment.

Unemployment

Unemployment is another urbanization problem. What is surprising is that a big percentage of youth who are unemployed belong to well raised and educated families. The job opportunities might be more in urban areas and also pay more, but as the number of people continues to grow, the jobs become even harder to find and retain.

Companies find themselves retrenching employees as well as putting even more people out of a job. All these factors put together to make the unemployment rate at an alarming high in urban areas. In recent times, with the Covid-19 outbreak, the unemployment rates have also increased manifolds. With people being laid off at an alarming rate, more and more people are becoming unemployed every day. Companies are retaining just the bare minimum number of employees.

It is alarming because, with the pandemic at our doors, we are facing the evils of urbanization even more. The jobs that had to be created once for fending the needs of the ever-growing urban population are being forced to shut down in the face of a huge economic crisis.

Housing Problems

Housing problems tend to develop when people move to cities and overcrowd in them. If the cities were not well prepared for the numbers, the houses become more scarce. It is even harder to settle people who come to cities and don't end up getting employed or those who settle in as immigrants. Some of these people are unable to afford to build their own homes or even paying rent.

The problem grows in intensity as the materials required to build new houses become more and more insufficient. Limited space also makes it difficult to facilitate construction, and financial resources may even be scarce as they become primarily channelled into other development and social safety net programs other than housing. This only adds to the problem.

Development of Slums

Urbanization and industrialization make a lot of people move to the urban areas, but they do not prepare them for the conditions they are likely to face when they arrive there. Urban areas tend to have a high cost of living. The housing problems fuel this even more as all the people who move to urban areas cannot be adequately accommodated.

This leads to the advent and growth of slums as safe havens for those who cannot afford the high costs of rent or lack substantial money to purchase apartments or build homes in urban areas. The slums arise from the construction of houses on under-developed or undervalued land due to how scarce and expensive apartments or land is in urban areas.

The houses in slums are often poorly constructed and most of the time lack basic amenities like clean water and proper sanitation. They are built to cater to low-income urban earners. Some of these settlements are even illegal and may be set up next to dumpsites, heavily polluted areas, or natural disaster risk areas such as swampy and mudflow areas.

Sanitation Problems

Sanitation problems are rampant in urban areas due to the overpopulation that is seen in many of the areas people settle. The local governments find it hard to properly set up and manage a proper sewerage system due to the rampant bulge of the human population. The fast increase in people's population sometimes overwhelms the local government's resource capacity to construct the required sanitation and sewage systems.

Sometimes, the existing sewerage systems may not have adequate human as well as the infrastructural capacity to treat and manage the waste. So instead, it is drained into the water sources – polluting them and endangering the health of urban dwellers who may not only contract harmful water-borne diseases but also consume water contaminated with industrial waste and heavy metals.

Water Shortage Problems

Water is very important for sustaining life. The rampant growth of the population in urban areas makes the water very scarce as the normal supply becomes strained and inadequate to properly meet the demands of the large population. What is more, the water problems may worsen with the increase in water pollution due to poor sewerage systems and a lack of preventive measures for managing local water pollution.

Health Hazards

People living in congested urban areas expose them to a lot of risks. The poor sanitation, water problems, and living in high-risk areas like next to dumpsites leads to disease of all kinds. To make matters worse, people in these areas often do not have proper access to health care services, which makes the disease much harder to cure, and at times, they even lead to death.

In the slums, people are often diagnosed with diseases like infertility, food poisoning, allergies, asthma, cardiovascular complication, respiratory failure, cancer, and death. This can all be accredited to the overall pollution that is experienced in these urban areas.

With the onset of the Covid-19 pandemic, the problem that global urbanization poses can be felt even more. Now that social distancing has become absolutely essential, urbanization has become the greatest health hazard of all times. In developing countries, especially, the problem is even more.

The slum areas are the worst affected. It is in these areas that the pandemic is spreading at a much higher rate. In most cases, the cases reported here are severe, and the fatality rate is very high. Here it is worth mentioning that, when the pandemic started spreading in India, people were really concerned about it spreading in Dharavi, India's largest slum situated in Mumbai. When it spread there, it was by sheer luck that it could be controlled at the earliest.

Degraded Environmental Quality

This is one of the most common effects of urbanization. The congestion of people in limited spaces and areas reduces the quality of air, contaminates water, and pollutes the noise and land. This leads to very poor environmental conditions for people to live and is often detrimental to the health of these people. There is also the need to improve the infrastructure so as to accommodate the rise in population by erecting new buildings and amenities.

This leads to the destruction of forest and natural habitats in order to acquire the materials required. Industrial waste poured to the river and lakes contaminates the water, and the noise brought about by the numerous human activities carried out sums up the many effects urbanizations bring in slum areas.

The more the urban population, the more pollution is caused by automobiles. Although carpooling is a relatively better practice, not everyone has come to terms with it. Most of the people prefer using their personal vehicles. Even if people opt for carpooling or public transport, it does not really help much.

Every day, millions of people avail of the commutes, which means, tens and thousands of vehicles run on the streets every single day. Most of these vehicles run on fossil fuels and, as a result, cause an immense amount of air pollution and degrades the quality of the air considerably.

Disposal of Trash

Urbanization has led to many factors that have made the trash disposal very difficult. The urban cities produce a lot of waste on a daily basis that they cannot properly dispose of this subjects the people living in these areas to multiple health risks. The areas that were initially set aside to accommodate trash disposal needs become full, and some of them are inhabited by slum people who move close to such areas.

Diseases are easily spread, with some often spread by the insects and animals that frequent the dumping site areas and then get into contact with water that people consume. The trash fills are also located in areas where people live which subjects them to many of these health hazards.

Transportation Problems

A lot of people are often moving around in between their workplaces and their homes; this more often leads to traffic jams and congestions. The number of people who own cars is growing every year, especially in urban areas and the public transport system is very unreliable.

The number of cars increases, and as a result of this, the traffic problems continue to worsen. This does not only lead to blockages in traffics but increases the chances of people getting involved in traffic accidents and urban air pollution.

Urban Crime

The more people are congested in urban areas, the higher the rate of unemployment as the available jobs are not enough to accommodate all. Resources have also become scarce, and not everyone has access to essential social services, which lead the disadvantaged to get into substance abuse, violence, burglary, and organized crime.

Lack of employment also increases poverty, which, as a result, makes it even harder for people to get the essential things they need to survive. People then turn to poverty-related crimes such as theft, conning, and organized crime as a way of earning a living.

Such criminal activities, mainly those rampant in urban areas due to poverty and lack of job opportunities, include kidnapping, robbery, carjacking, rape, and even murder. These acts make cities very hard for people to live in as they are not guaranteed their safety, especially for victims who cannot defend themselves in such situations.

Increased Rates of Poverty

Global urbanization ultimately leads to poverty. As the rates of unemployment increases, more and more people continue to sink beneath the poverty line. Also, as the population increases and the urban areas become more and more congested, the state government starts losing track of the population.

As it does so, in some cases, it also fails to provide for all the people adequately. It ultimately leads to extreme forms of poverty where people have to live on the streets with only a little or nothing to eat or to drink.

As a result of poverty, the standard of living of people also decreases to a point where it can also sometimes be rendered as inhuman. In a world where the rich are constantly becoming richer, and the poor are constantly becoming poorer, poverty is by far one of the largest threats to human existence.

References

- Urbanization-and-urban-growth: conserve-energy-future.com, Retrieved 21, May 2020
- Urbanization-problems: conserve-energy-future.com, Retrieved 15, July 2020

3

Urban Structure, Planning and Design

The arrangement of land use in urban areas is known as urban structure. The process which is involved in developing and designing built environment and land use in urban areas is called urban planning. The topics elaborated in this chapter will help in gaining a better perspective about the different aspects of urban structure, planning and design.

Urban Structure

Urban structure is the arrangement of land use in urban areas. Sociologists, economists, and geographers have developed several models, explaining where different types of people and businesses tend to exist within the urban setting. Urban structure can also refer to the urban spatial structure, which concerns the arrangement of public and private space in cities and the degree of connectivity and accessibility.

Zonal Model

This model was the first to explain distribution of social groups within urban areas. Based on one single city, Chicago, it was created by sociologist Ernest Burgess in 1924. According to this model, a city grows outward from a central point in a series of rings. The innermost ring represents the central business district. It is surrounded by a second ring, the zone of transition, which contains industry and poorer-quality housing. The third ring contains housing for the working-class and is called the zone of independent workers' homes. The fourth ring has newer and larger houses usually occupied by the middle-class. This ring is called the zone of better residences. The outermost ring is called the commuter's zone. This zone represents people who choose to live in residential suburbs and take a daily commute into the CBD to work.

Sectoral Model

A second theory of urban structure was proposed in 1939 by an economist named Homer Hoyt. His model, the sector model, proposed that a city develops in sectors instead of rings. Certain areas of a city are more attractive for various activities, whether by chance or geographic and environmental reasons. As the city grows and these activities flourish and expand outward, they do so in a wedge and become a sector of the city. If a district is set up for high income housing, for example, any new development in that district will expand from the outer edge.

To some degree this theory is just a refinement on the concentric model rather than a radical restatement. Both Hoyt and Burgess claimed Chicago supported their model. Burgess claimed that Chicago's central business district was surrounded by a series of rings, broken only by Lake Michigan. Hoyt argued that the best housing developed north from the central business district along Lake Michigan, while industry located along major rail lines and roads to the south, southwest, and northwest.

Multiple Nuclei Model

Geographers Chauncy Harris and Edward Ullman developed the multiple nuclei model in 1945. According to this model, a city contains more than one center around which activities revolve. Some activities are attracted to particular nodes while others try to avoid them. For example, a university node may attract well-educated residents, pizzerias, and bookstores, whereas an airport may attract hotels and warehouses. Other businesses may also form clusters, sometimes known locally as Iron Triangles for automobile repair or red light districts for prostitution, or arts districts. Incompatible activities will avoid clustering in the same area, explaining why heavy industry and high-income housing rarely exist in the same neighbourhood.

Urban Activity System

Many observers of urban areas have pointed out that bulk of urban activities occurs in ordered hierarchies. Certain activities, such as elementary schools and small neighborhood shopping centers, exist at a scale that serves only local areas. Other type of urban activities exists in large concentrations and may serve an entire urban region. Examples of region wide activities include major cultural facilities, Post secondary education facilities, specialized retail trade outlets, and certain types of personal service.

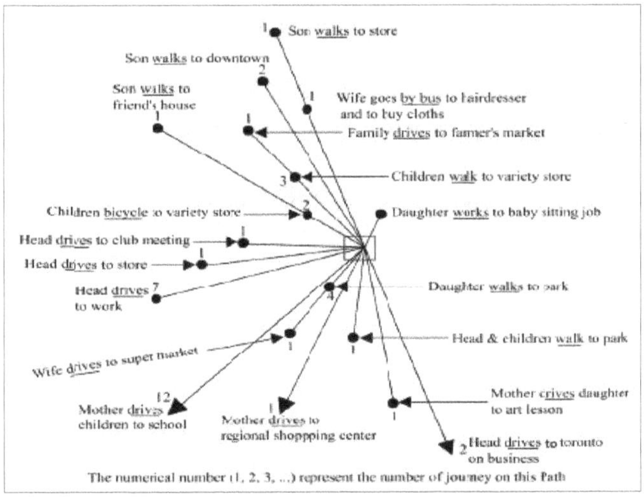

Weekly household activity Pattern.

The above figure shows the weekly activity pattern for the members of one household. This diagram indicates approximately the length and location of each trip, the mode of travel used, and the number of trips made throughout the week.

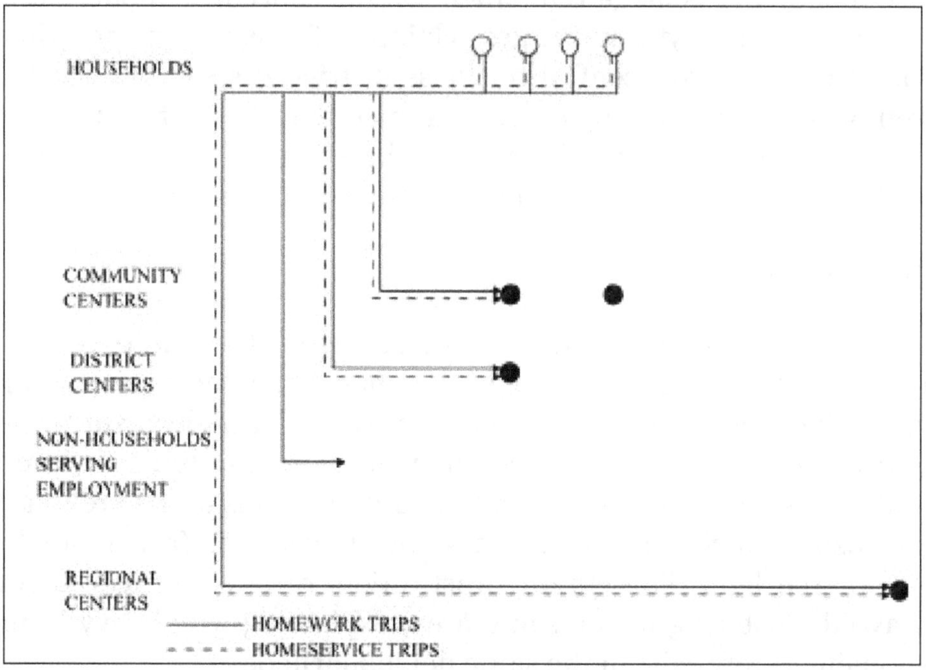

Hierarchical structure of household – based activities.

The above figure illustrates an idealization of the hierarchy of some of the activities located in urban areas. Two activity systems are identified in the figure and these are the household- workplace subsystem and the household service place sub system. Service places are of a very general nature and include school, personal services, and certain types of cultural and environment facilities. The activities labeled as 'non-household-serving employment' are-

(1) Employment in basic industries.

(2) Employment in population serving industries.

The regional center might contain government offices specialized retail trade outlets and personal service as well as the major cultural facility of the region

NB: - Home-based travel demand represents 80% of the total travel in urban areas. In addition, travel by commercial vehicles in normally 12 to 15% of total vehicle trips with in urban areas.

Urban Movement Hierarchies

Transport Systems Provide two fundamental functions and these movement and access to land based activities. There is a conflict between these two functions.

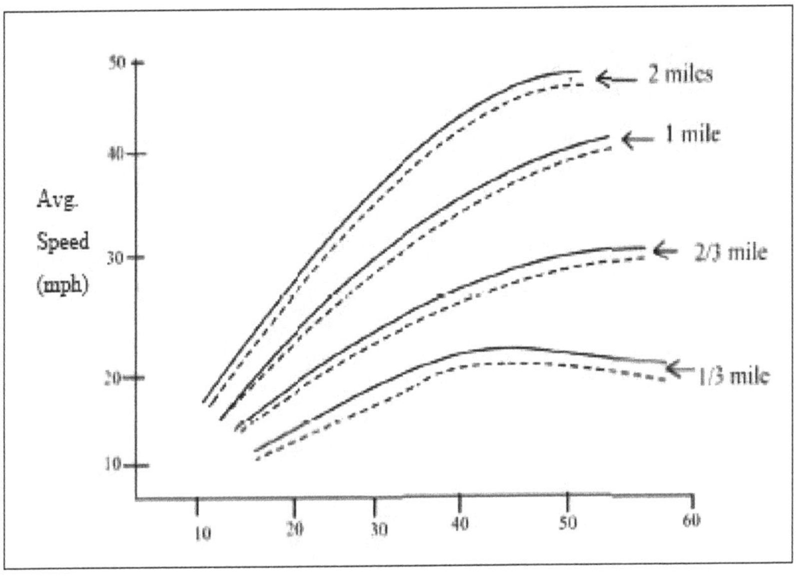

Top speed of vehicle (mph).

Average Speed of travel by mass transit vs top Vehicle Speed, Station Spacing and Station Stop Time.

The above figure shows that the station spacing on rapid transit lines controlling factor in the average speed of travel on these lines. In other words, as access to a rapid transit line is increased through decreased station spacing, the efficiency of movement decreases. A Similar effect for road travel has been illustrated that the efficiency of travel along controlled access roads is far superior to travel along arterial and collector roads without control of access, or with limited control of access.

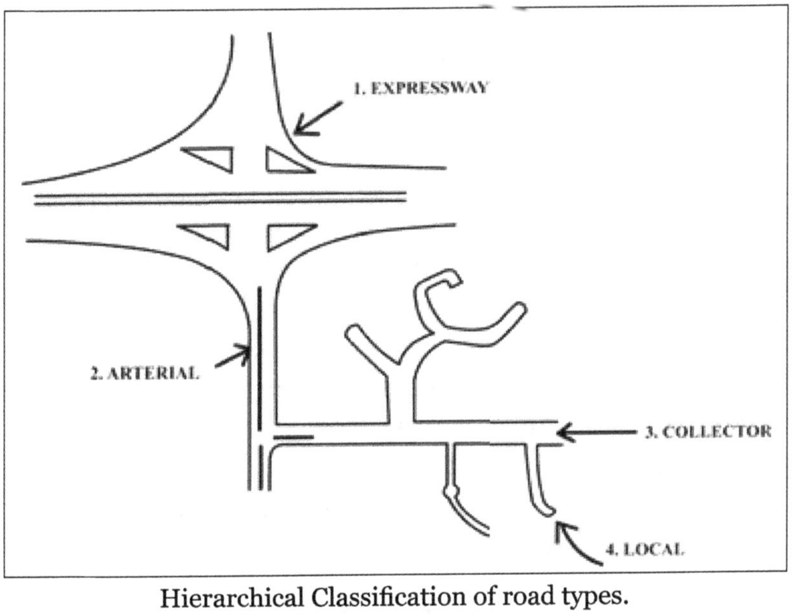

Hierarchical Classification of road types.

The above figure illustrates one hierarchical Classification of roads which consists of four classes with the following Properties.

Urban Planning

Urban planning designs settlements, from the smallest towns to the largest cities. Shown here is Hong Kong from Western District overlooking Kowloon, across Victoria Harbour.

Partizánske in Slovakia – an example of a typical planned European industrial city founded in 1938 together with a shoemaking factory in which practically all adult inhabitants of the city were employed.

Urban planning is a technical and political process concerned with the development and use of land, planning permission, protection and use of the environment, public welfare, and the design of the urban environment, including air, water, and the infrastructure passing into and out of urban areas, such as transportation, communications, and distribution networks. Urban planning is also referred to as urban and regional planning, regional planning, town planning, city planning, rural planning or some combination in various areas worldwide. It takes many forms and it can share perspectives and practices with urban design.

Urban planning guides orderly development in urban, suburban and rural areas. Although predominantly concerned with the planning of settlements and communities, urban planning is also responsible for the planning and development of water use and resources, rural and agricultural land, parks and conserving areas of natural environmental significance. Practitioners of urban planning are concerned with research and analysis, strategic thinking, architecture, urban design, public consultation, policy recommendations, implementation and management.

Urban planners work with the cognate fields of architecture, landscape architecture, civil engineering, and public administration to achieve strategic, policy and sustainability goals. Early urban planners were often members of these cognate fields. Today urban planning is a separate, independent professional discipline. The discipline is the broader category that includes different sub-fields such as land-use planning, zoning, economic development, environmental planning, and transportation planning.

History

There is evidence of urban planning and designed communities dating back to the Mesopotamian, Indus Valley, Minoan, and Egyptian civilizations in the third millennium BCE. Archeologists studying the ruins of cities in these areas find paved streets that were laid out at right angles in a grid pattern. The idea of a planned out urban area evolved as different civilizations adopted it. Beginning in the 8th century BCE, Greek city states were primarily centered on orthogonal (or grid-like) plans. The ancient Romans, inspired by the Greeks, also used orthogonal plans for their cities. City planning in the Roman world was developed for military defense and public convenience. The spread of the Roman Empire subsequently spread the ideas of urban planning. As the Roman Empire declined, these ideas slowly disappeared. However, many cities in Europe still held onto the planned Roman city center. Cities in Europe from the 9th to 14th centuries, often grew organically and sometimes chaotically. But many hundreds of new towns were newly built according to preconceived plans, and many others were enlarged with newly planned extensions. Most of these were realized from the 12th to 14th centuries, with a peak-period at the end of the 13th. From the 15th century on, much more is recorded of urban design and the people that were involved. In this period, theoretical treatises on architecture and urban planning start to appear in which theoretical questions are addressed and designs of towns and cities are described and depicted. During the Enlightenment period, several European rulers ambitiously attempted to redesign capital cities. During the Second French Republic, Baron Georges-Eugène Haussmann, under the direction of Napoleon III, redesigned the city of Paris into a more modern capital, with long, straight, wide boulevards.

Planning and architecture went through a paradigm shift at the turn of the 20th century. The industrialized cities of the 19th century grew at a tremendous rate. The pace and style of this industrial construction was largely dictated by the concerns of private business. The evils of urban life for the working poor were becoming increasingly evident as a matter for public concern. The laissez-faire style of government management of the economy, in fashion for most of the Victorian era, was starting to give way to a

New Liberalism that championed intervention on the part of the poor and disadvantaged. Around 1900, theorists began developing urban planning models to mitigate the consequences of the industrial age, by providing citizens, especially factory workers, with healthier environments.

Urban planning started to become professionalized during this time. The Town and Country Planning Association was founded in 1899 and the first academic course in Great Britain on urban planning was offered by the University of Liverpool in 1909. In the 1920s, the ideas of modernism and uniformity began to surface in urban planning, and lasted until the 1970s. Many planners started to believe that the ideas of modernism in urban planning led to higher crime rates and social problems. Urban planners now focus more on individualism and diversity in urban centers.

Theories

Planning theory is the body of scientific concepts, definitions, behavioral relationships, and assumptions that define the body of knowledge of urban planning. There are eight procedural theories of planning that remain the principal theories of planning procedure today: the rational-comprehensive approach, the incremental approach, the transactive approach, the communicative approach, the advocacy approach, the equity approach, the radical approach, and the humanist or phenomenological approach.

Technical Aspects

Technical aspects of urban planning involve the applying scientific, technical processes, considerations and features that are involved in planning for land use, urban design, natural resources, transportation, and infrastructure. Urban planning includes techniques such as: predicting population growth, zoning, geographic mapping and analysis, analyzing park space, surveying the water supply, identifying transportation patterns, recognizing food supply demands, allocating healthcare and social services, and analyzing the impact of land use.

Urban Planners

An urban planner is a professional who works in the field of urban planning for the purpose of optimizing the effectiveness of a community's land use and infrastructure. They formulate plans for the development and management of urban and suburban areas, typically analyzing land use compatibility as well as economic, environmental and social trends. In developing the plan for a community (whether commercial, residential, agricultural, natural or recreational), urban planners must also consider a wide array of issues such as sustainability, air pollution, traffic congestion, crime, land values, legislation and zoning codes.

The importance of the urban planner is increasing throughout the 21st century, as modern society begins to face issues of increased population growth, climate change and unsustainable development. An urban planner could be considered a green collar professional.

Urban Design

Urban Design Center Kashiwa-no-ha.

Urban design is the process of designing and shaping cities, towns and villages. In contrast to architecture, which focuses on the design of individual buildings, urban design deals with the larger scale of groups of buildings, streets and public spaces, whole neighborhoods and districts, and entire cities, with the goal of making urban areas functional, attractive, and sustainable.

Urban design is an inter-disciplinary subject that utilizes elements of many built environment professions, including landscape architecture, urban planning, architecture, civil and municipal engineering. It is common for professionals in all these disciplines to practice in urban design. In more recent times different sub-strands of urban design have emerged such as strategic urban design, landscape urbanism, water-sensitive urban design, and sustainable urbanism.

Urban design demands an understanding of a wide range of subjects from physical geography, through to social science, and an appreciation for disciplines, such as real estate development, urban economics, political economy and social theory.

Urban design is about making connections between people and places, movement and urban form, nature and the built fabric. Urban design draws together the many strands of place-making, environmental stewardship, social equity and economic viability into the creation of places with distinct beauty and identity. Urban design draws these and other strands together creating a vision for an area and then deploying the resources and skills needed to bring the vision to life.

Urban design theory deals primarily with the design and management of public space (i.e. the 'public environment', 'public realm' or 'public domain'), and the way public places are experienced and used. Public space includes the totality of spaces used freely on a day-to-day basis by the general public, such as streets, plazas, parks and public infrastructure. Some aspects of privately owned spaces, such as building facades or domestic gardens, also contribute to public space and are therefore also considered by urban design theory. Important writers on urban design theory include Christopher Alexander, Peter Calthorpe, Gordon Cullen, Andres Duany, Jane Jacobs, Mitchell Joachim, Jan Gehl, Allan B. Jacobs, Kevin Lynch, Aldo Rossi, Colin Rowe, Robert Venturi, William H. Whyte, Camillo Sitte, Bill Hillier (Space syntax) and Elizabeth Plater-Zyberk.

History

Although contemporary professional use of the term 'urban design' dates from the mid-20th century, urban design as such has been practiced throughout history. Ancient examples of carefully planned and designed cities exist in Asia, Africa, Europe and the Americas, and are particularly well-known within Classical Chinese, Roman and Greek cultures (see Hippodamus of Miletus).

European Medieval cities are often, and often erroneously, regarded as exemplars of undesigned or 'organic' city development. There are many examples of considered urban design in the Middle Ages (see, e.g., David Friedman, *Florentine New Towns: Urban Design in the Late Middle Ages*, MIT 1988). In England, many of the towns listed in the 9th century Burghal Hidage were designed on a grid, examples including Southampton, Wareham, Dorset and Wallingford, Oxfordshire, having been rapidly created to provide a defensive network against Danish invaders. 12th century western Europe brought renewed focus on urbanisation as a means of stimulating economic growth and generating revenue. The burgage system dating from that time and its associated burgage plots brought a form of self-organising design to medieval towns. Rectangular grids were used in the Bastides of 13th and 14th century Gascony, and the new towns of England created in the same period.

Throughout history, design of streets and deliberate configuration of public spaces with buildings have reflected contemporaneous social norms or philosophical and religious beliefs (see, e.g., Erwin Panofsky, *Gothic Architecture and Scholasticism*, Meridian Books, 1957). Yet the link between designed urban space and human mind appears to be bidirectional. Indeed, the reverse impact of urban structure upon human behaviour and upon thought is evidenced by both observational study and historical record. There are clear indications of impact through Renaissance urban design on the thought of Johannes Kepler and Galileo Galilei (see, e.g., Abraham Akkerman, "Urban planning in the founding of Cartesian thought," *Philosophy and Geography* 4(1), 2001). Already René Descartes in his *Discourse on the Method* had attested to the impact Renaissance planned new towns had upon his own thought, and much evidence exists that the Renaissance streetscape was also the perceptual stimulus that had led to the development of coordinate geometry.

The beginnings of modern urban design in Europe are associated with the Renaissance but, especially, with the Age of Enlightenment. Spanish colonial cities were often planned, as were some towns settled by other imperial cultures. These sometimes embodied utopian ambitions as well as aims for functionality and good governance, as with James Oglethorpe's plan for Savannah, Georgia. In the Baroque period the design approaches developed in French formal gardens such as Versailles were extended into urban development and redevelopment. In this period, when modern professional specialisations did not exist, urban design was undertaken by people with skills in areas as diverse as sculpture, architecture, garden design, surveying, astronomy, and military engineering. In the 18th and 19th centuries, urban design was perhaps most closely linked with surveyors (engineers) and architects. The increase in urban populations brought with it problems of epidemic disease, the response to which was a focus on public health, the rise in the UK of municipal engineering and the inclusion in British legislation of provisions such as minimum widths of street in relation to heights of buildings in order to ensure adequate light and ventilation.

Much of Frederick Law Olmsted's work was concerned with urban design, and the newly formed profession of landscape architecture also began to play a significant role in the late 19th century.

Modern Urban Design

Modern urban design is a part of the wider discipline of Urban planning. Indeed, Urban planning began as a movement primarily occupied with matters of public health and urban design.

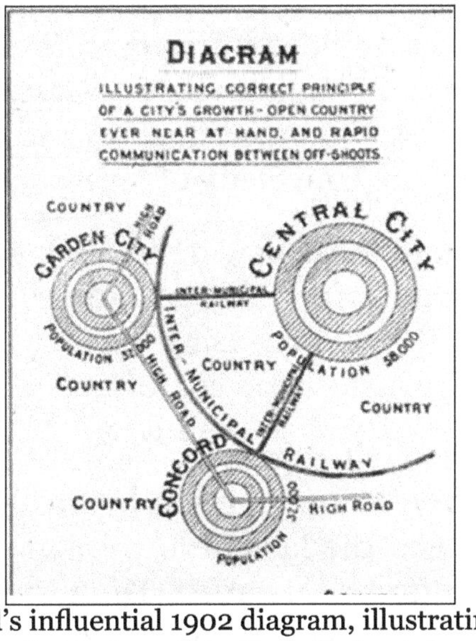

Ebenezer Howard's influential 1902 diagram, illustrating urban growth through garden city "off-shoots".

Planning and architecture went through a paradigm shift at the turn of the 20th century. The industrialized cities of the 19th century had grown at a tremendous rate, with the pace and style of building largely dictated by private business concerns. The evils of urban life for the working poor were becoming increasingly evident as a matter for public concern. The laissez-faire style of government management of the economy, in fashion for most of the Victorian era, was starting to give way to a New Liberalism that championed intervention on the part of the poor and disadvantaged. Around 1900, theorists began developing urban planning models to mitigate the consequences of the industrial age, by providing citizens, especially factory workers, with healthier environments.

The first modern urban planning theorist was Sir Ebenezer Howard, who initiated the garden city movement in 1898. Howard's ideas, although utopian were also highly practical and were adopted around the world in the ensuing decades. His garden cities were intended to be planned, self-contained communities surrounded by parks, containing proportionate and separate areas of residences, industry and agriculture. Inspired by the Utopian novel *Looking Backward* and Henry George's work *Progress and Poverty*, Howard published his book *Garden Cities of To-morrow* in 1898, commonly regarded as the most important book in the history of urban planning. His idealised garden city would house 32,000 people on a site of 6,000 acres (2,428 ha), planned on a concentric pattern with open spaces, public parks and six radial boulevards, 120 ft (37 m) wide, extending from the centre. The garden city would be self-sufficient and when it reached full population, another garden city would be developed nearby. Howard envisaged a cluster of several garden cities as satellites of a central city of 50,000 people, linked by road and rail.

The first garden cities were created at Letchworth and Welwyn Garden City in Hertfordshire. Sir Frederic Osborn extended the movement to regional planning.

Urban planning became professionalized at this period, with input from utopian visionaries as well as from the practical minded infrastructure engineers and local councilors combining to produce new design templates for political consideration. The Town and Country Planning Association was founded in 1899 and the first academic course on urban planning was offered by the University of Liverpool in 1909.

The first official consideration of these new trends was embodied in the Housing and Town Planning Act of 1909 that compelled local authorities to introduce coherent systems of town planning across the country using the new principles of the 'garden city', and to ensure that all housing construction conformed to specific building standards.

Following this Act, surveyors, civil engineers, architects, lawyers and others began working together within local government in the United Kingdom to draw up schemes for the development of land and the idea of town planning as a new and distinctive area of expertise began to be formed. In 1910, Thomas Adams was appointed as the first Town Planning Inspector at the Local Government Board, and began meeting with practitioners. The Town Planning Institute was established in 1914 with a mandate to advance the study of town-planning and civic design. The first university course in America was established at Harvard University in 1924.

The automobile was an important influence on the design of urban development in the 20th century, and the rise of the "urban design" movement can be seen in part as a reaction to the adverse impact of car-use and car orientated design. 'Urban design' was first used as a distinctive term when Harvard University hosted a series of Urban Design Conferences from 1956 . These conferences provided a platform for the launching of Harvard's Urban Design program in 1959-60. The writings of Jane Jacobs, Kevin Lynch, Gordon Cullen and Christopher Alexander became authoritative works for the school of Urban Design.

Gordon Cullen's *The Concise Townscape,* first published in 1961, also had a great influence on many urban designers. Cullen examined the traditional artistic approach to city design of theorists such as Camillo Sitte, Barry Parker and Raymond Unwin. He created the concept of 'serial vision', defining the urban landscape as a series of related spaces.

Jane Jacobs' *The Death and Life of Great American Cities*, published in 1961, was also a catalyst for interest in ideas of urban design. She critiqued the Modernism of CIAM, and asserted that the publicly unowned spaces created by the 'city in the park' notion of Modernists was one of the main reasons for the rising crime rate. She argued instead for an 'eyes on the street' approach to town planning, and the resurrection of main public space precedents, such as streets and squares, in the design of cities.

Kevin Lynch's *The Image of the City* of 1961 was also seminal to the movement, particularly with regards to the concept of legibility, and the reduction of urban design theory to five basic elements - paths, districts, edges, nodes, landmarks. He also made popular the use of mental maps to understanding the city, rather than the two-dimensional physical master plans of the previous 50 years.

Other notable works include Rossi's *Architecture of the City* (1966), Venturi's *Learning from Las Vegas* (1972), Colin Rowe's *Collage City* (1978), and Peter Calthorpe's *The Next American Metropolis* (1993). Rossi introduced the concepts of 'historicism' and 'collective memory' to urban design, and proposed a 'collage metaphor' to understand the collage of new and older forms within the same urban space. Calthorpe, on the other hand, developed a manifesto for sustainable urban living via medium density living, as well as a design manual for building new settlements in accordance with his concept of Transit Oriented Development (TOD). Bill Hillier and Julienne Hanson in "The Social Logic of Space" (1984) introduced the concept of Space Syntax to predict how movement patterns in cities would contribute to urban vitality, anti-social behaviour and economic success. The popularity of these works resulted in terms such as 'historicism', 'sustainability', 'livability', 'high quality of urban components', etc. become everyday language in the field of urban planning.

Current Trends

Jakriborg in Sweden, started in the late 1990s as a new urbanist eco-friendly new town near Malmö.

Various current movements in urban design seek to create sustainable urban environments with long-lasting structures, buildings and a great liveability for its inhabitants. The most clearly defined form of walkable urbanism is known as the *Charter of New Urbanism*. It is an approach for successfully reducing environmental impacts by altering the built environment to create and preserve smart cities which support sustainable transport. Residents in compact urban neighborhoods drive fewer miles, and have significantly lower environmental impacts across a range of measures, compared with those living in sprawling suburbs. The concept of Circular flow land use management has also been introduced in Europe to promote sustainable land use patterns, that strive for compact cities and a reduction of greenfield land taken by urban sprawl.

In sustainable construction the recent movement of New Classical Architecture promotes a sustainable approach towards urban construction that appreciates and develops smart growth, walkability, architectural tradition and classical design. This in contrast to modernist and globally uniform architecture, as well as opposing solitary housing estates and suburban sprawl. Both trends started in the 1980s.

Principles

L'Enfant's plan for Washington DC.

Gehl Architects' project for Brighton New Road employing shared space.

Public spaces are frequently subject to overlapping management responsibilities of multiple public agencies or authorities and the interests of nearby property owners, as well as the requirements of multiple and sometimes competing users. The design, construction and management of public spaces therefore typically demands consultation and negotiation across a variety of spheres. Urban designers rarely have the degree of artistic liberty or control sometimes offered in design professions such as architecture. It also typically requires interdisciplinary input with balanced representation of multiple fields including engineering, ecology, local history, and transport planning.

The scale and degree of detail considered varies depending on context and needs. It ranges from the layout of entire city regions, cities, as with l'Enfant's plan for Washington DC, Griffin and Mahony's plan for Canberra and Doxiadis' plan for Islamabad (although such opportunities are obviously rare), through 'managing the sense of a region' as described by Kevin Lynch, to the design of street furniture.

Urban design may encompass the preparation of design guidelines and regulatory frameworks, or even legislation to control development, advertising, etc. and in this sense overlaps with urban planning. It may encompass the design of particular spaces and structures and in this sense overlaps with architecture, landscape architecture, highway engineering and industrial design. It may also deal with 'place management' to guide and assist the use and maintenance of urban areas and public spaces.

Most urban design work is undertaken by urban planners, landscape architects and architects. There are also professionals who identify themselves specifically as urban designers. Many architecture, landscape and planning programs incorporate urban design theory and design subjects into their curricula and there are an increasing number of university programs offering degrees in urban design, usually at post-graduate level.

Urban Design Considers:

- Pedestrian zones.
- Incorporation of nature within a city.
- Aesthetics.
- *Urban structure* – How a place is put together and how its parts relate to each other.
- *Urban typology, density* and sustainability - spatial types and morphologies related to intensity of use, consumption of resources and production and maintenance of viable communities.
- *Accessibility* – Providing for ease, safety and choice when moving to and through places.
- *Legibility and wayfinding* – Helping people to find their way around and understand how a place works.
- *Animation* – Designing places to stimulate public activity.
- *Function and fit* – Shaping places to support their varied intended uses.
- *Complementary mixed uses* – Locating activities to allow constructive interaction between them.
- *Character and meaning* – Recognizing and valuing the differences between one place and another.
- *Order and incident* – Balancing consistency and variety in the urban environment in the interests of appreciating both.
- *Continuity and change* – Locating people in time and place, including respect for heritage and support for contemporary culture.
- *Civil society* – Making places where people are free to encounter each other as civic equals, an important component in building social capital.

Equality Issues

Until the 1970s, the design of towns and cities took little account of the needs of people with disabilities. At that time, disabled people began to form movements demanding recognition of their potential contribution if social obstacles were removed. Disabled people challenged the 'medical model' of disability which saw physical and mental problems as an individual 'tragedy' and people with disabilities as 'brave' for enduring them. They proposed instead a 'social model' which said that barriers to disabled people result from the design of the built environment and attitudes of able-bodied people. 'Access Groups' were established composed of people with disabilities who audited their local areas, checked planning applications and made representations for improvements. The new profession of 'access officer' was established around that time to produce guidelines based on the recommendations of access groups and to oversee adaptations to existing buildings as well as to check on the accessibility of new proposals. Many local authorities now employ access officers who are regulated by the Access Association. A new chapter of the Building Regulations (Part M) was introduced in 1992. Although it was beneficial to have legislation on this issue the requirements were fairly minimal but continue to be improved with ongoing amendments. The Disability Discrimination Act 1995 continues to raise awareness and enforce action on disability issues in the urban environment.

Urban Roads

The roads are named according to the type of constructions, jurisdiction and important function etc. Names like earth, metalled roads, asphalt roads, and concrete roads indicate the type of constructions. Names like local roads, district roads, state highways, national highways indicate their jurisdiction. Names like rectangular roads, ring roads, and diagonal roads, radial and circular roads indicate their geometric shape. Names like Avenue, Promenade (a pleasure drive, with water front at least on side), Boulevards and Parkways indicate their dominant function.

The urban roads are classified as per their importance such as:

I. Arterial roads.

II. Secondary or sub-arterial roads.

III. Local roads.

IV. Other roads.

Arterial Roads

These are that roads which connect the town to state highway or a national highway. They pass through the city limits and carry a large amount of traffic and therefore should be planned as straight as possible, avoiding sharp curves. Change in direction should be a accompanied by smooth curves. These should not enter into the heart of the city at any cost, should have very few road junctions, which should be controlled by roundabouts or fly-overs. They should have no obstructions such as frontage of buildings, loading or unloading areas, parking places, and pedestrians on the carriage way. Further these roads may be made more pleasing creating squares, erecting public and semi- public buildings at the focal points to invite through traffic and encourage speedy transportation by removing all types of traffic barriers. The width of these roads should not be less than 25 m. to 30 m.

Secondary or Sub-arterial Roads

Also known as major roads they run within the limits of the town connecting its important centres. They are designed for slow moving traffic and cover a short distance. The sub- arterial roads act as a link between the arterial roads and local roads. The sub-arterial roads should be improved and provided with safety measures at intersections.

Local Roads

These roads, also known as minor roads, are meant to provide approach to the buildings, officers, shops, schools, colleges etc. There should be no through traffic here and so the local roads are not linked with the arterial roads. These roads need not be straight but can follow the contours of the land.

These roads are used for residential units, shopping and business centres. They therefore form the pocket or precinct roads mainly to serve the non-vehicular traffic. The width of these roads should not be less than 7m to 10m.

Other Roads

(a) By-pass Roads

When the main or through roads pass through the congested areas of the towns, there will be considerable reduction in the speed of the vehicles and the smooth flow of the traffic is largely affected resulting loss of time and fuel. In order to maintain the smooth and speedy flow of traffic, bypass roads are constructed. These are also called as loop-roads through which the main traffic can pass from one side and again join on the main road on the other side thus avoiding the congested area or ribbon development of the town.

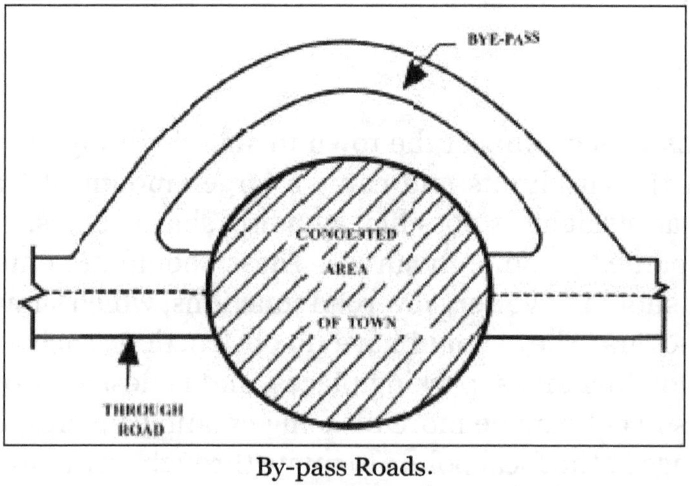

By-pass Roads.

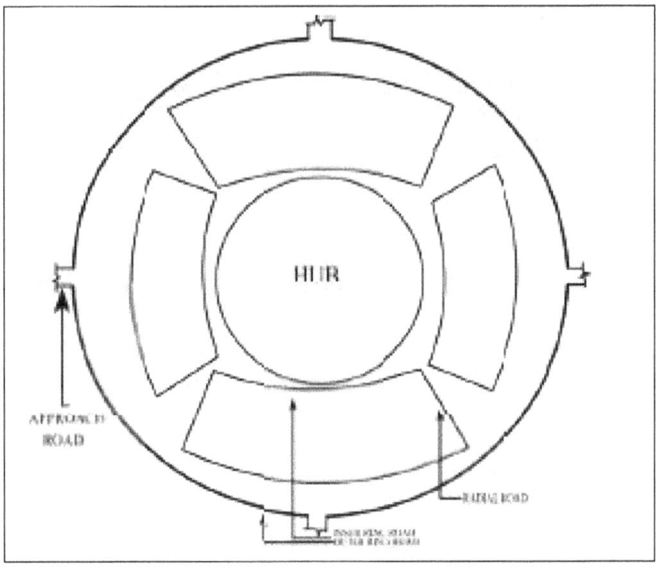

Outer and Inner Ring Roads.

(b) Outer and inner Ring Roads

These roads are in the form of circles or rings and hence the name. The outer ring road is meant to divert the through traffic approaching the town. The inner ring road is meant to divert the local from through traffic. These ring roads help to reduce the traffic congestion of the large towns.

(c) Express-ways

The express-way is meant to function as arterial road for the movement of fast moving traffic in the big metropolitan cities like, Bombay, Calcutta, Delhi etc. Two to three such express-ways are necessarily to be provided around big cities of modern days to face the tremendous growth of the traffic. They however should not form a part of the regular street system, although they should be suitably joined and linked with them. Express-ways are designed with easy gradients and smooth curves so as to carry the traffic speedily and safely. These are originated from the German Autobahnen and Italian Autostrade. These are comparable next to railways in cost and carrying capacity of traffic.

(d) Free-ways

These are the special routes meant to carry fast moving traffic and therefore designed with high standard of alignment, clear visibility, wide carriage way, easy gradient and smooth curves etc.

There is no access from adjacent properties as a result full width of free-way is made available for the fast moving vehicles without any obstruction. The free-ways function as arterial roads passing around the city with controlled access. They also act as main entrances and exits as such they form a part of major road system.

Types of Urban Road System

Following are the different system of streets or roads commonly adopted in the city:

- Rectangular or Grid-Iron Street system.
- Rectangular combined with Diagonal street system.
- Concentric and Radial Street system.
- Rectangular combined with Radial street system.
- Organic street system.
- Irregular Medieval street system.
- Combination of Rectangular and Irregular Street system.

Rectangular or Grid-iron Street System

This system is suitable for a fairly plain country, without any predominant natural features. Here the streets have equal widths and they cross each other at right angles.

The advantages of this system are

- It is convenient to traffic and so a speedy and free traffic can be maintained.
- The houses are constructed in rectangular blocks so convenient, economical and most suited for building construction.
- There is no wastage of land since no irregular portions are left out.
- The maximum area is used for construction. It is therefore most advantageous for private land-owners.

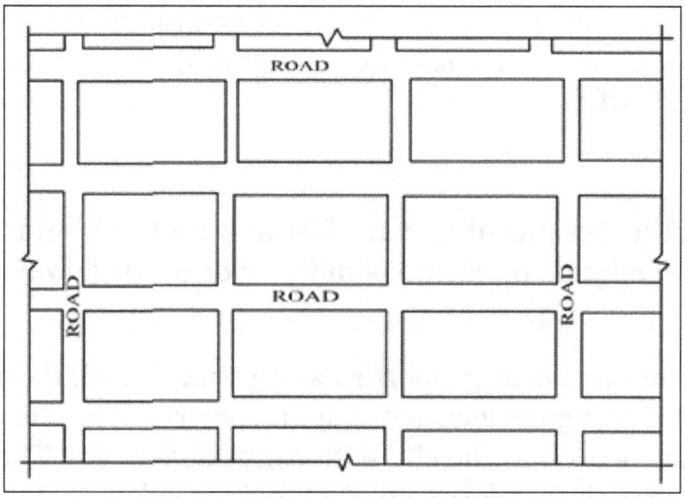

Rectangular or Grid-Iron System.

The disadvantages of this system are:

I. This system does not provide short cuts which provide a direct access to trade and shopping centre.

II. In uneven country, this system leads to inconvenience, discomfort and moreover becomes expensive.

III. This system has too many junctions and crossings. So there are more chances for road accidents. Hence it is most unsatisfactory from traffic point of view.

IV. This system is too mathematical and produces monotonous effect since the roads are s raight as a result the vistas are open, devoid of interest and variety.

Rectangular Combined with Diagonal Street System

It is an improved type of rectangular system. England was the first town which was designed as this type of system.

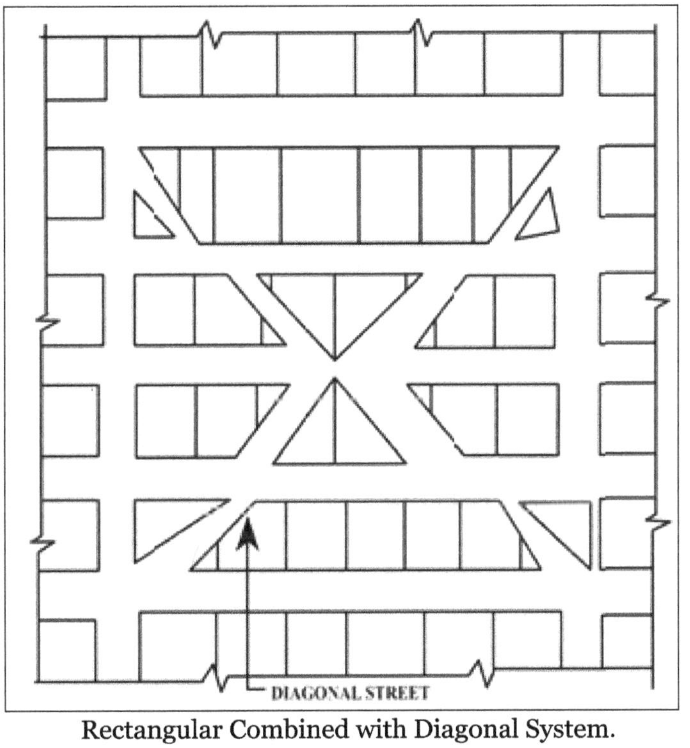

Rectangular Combined with Diagonal System.

The advantages of the system are:

I. This system rovides direct communication from distant parts of the town.

II. At the place where the diagonals meet, a park, garden, fountain or memorial can be up to add aesthetics to the town.

III. This system is most practicable as there are very few obstacles.

The disadvantages are:

This system gives rise to many dangerous intersections at the junctions of diagonal streets and rectangular streets and so most unsatisfactory from traffic point of view E.g. Washington, the Capital of U.S.A.

Concentric and Radial Street System

In this system the circular or ring road are connect to radial roads. It is also called 'Spider's Web System'. Figure (a) and (b).

This system represents the most natural growth as many cities roughly possess a part of this feature. Here the town grows in the form of concentric Ring roads round the hub of the town so that each growth r part is as near to the centre. E.g. Vienna, Austria.

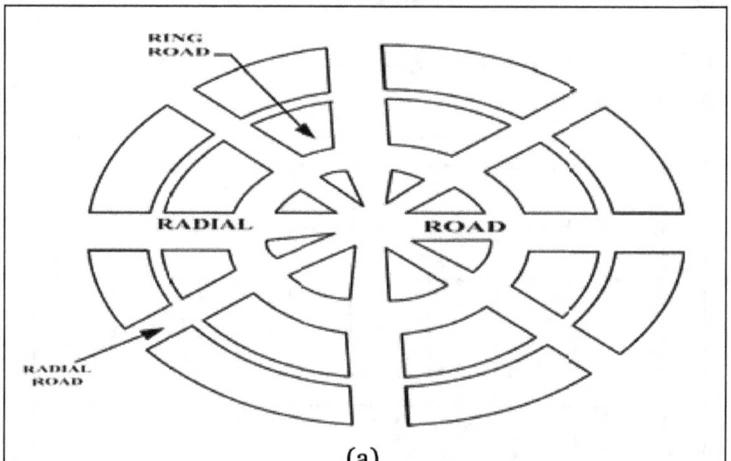

(a)

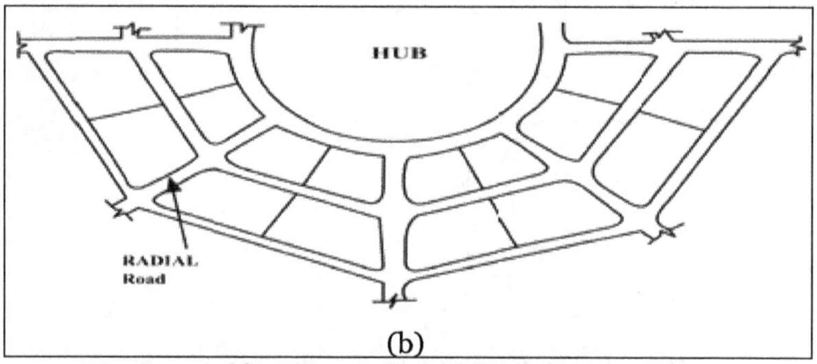

(b)

Figure a and b: Concentric and Radial street system.

The advantages of the system are:

I. The radial roads provide direct access to the heart or central part of the town and are earmarked for fast moving traffic. Hence, these ring roads can be designed as arterial roads.

II. The circular roads function as the intermediary between diagonal roads and local roads.

III. The circular roads help to distribute the local traffic to the different thoroughfares. Hence it is most suitable from traffic point of view.

IV. This system is found to be more useful for the economic expansion of the town particularly if the town has central features such as important public buildings, market etc.

The disadvantages of this system are:

I. The plots are formed trapezoidal in shape. So a lot of space is wasted due to irregular portions left out.

II. It is not economical for the construction of houses, since maximum area is not utilized.

Rectangular Combined with Radial Street System

In practice a combination of rectangular and radial system can be adopted if the topographical features are favourable to suit the local requirements of each p ace. It has both the advantages of rectangular roads and radial roads.

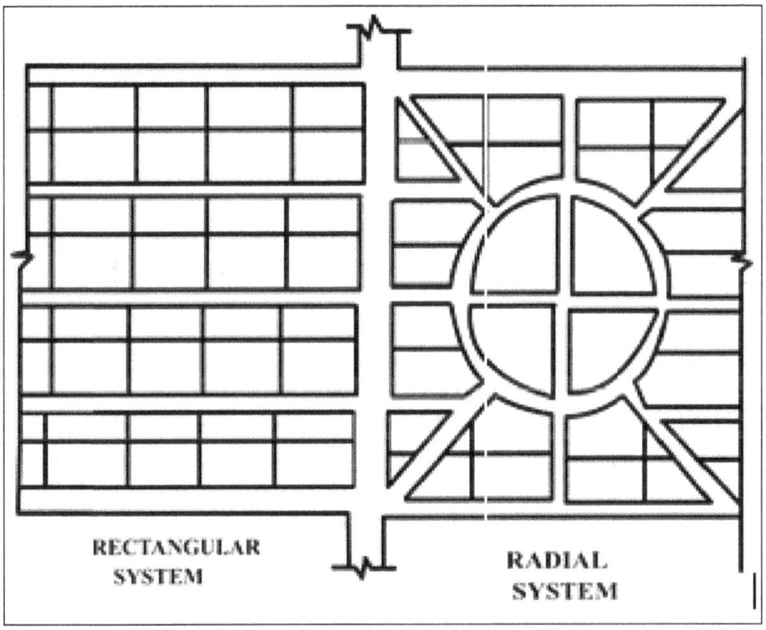

Rectangular combined with Radial street system.

Organic Street System

It is also called as Topographical and Informal Street system. The name 'Organic' is applied because the plan resembles to a microscope slide, presenting the cell structure of animal or vegetable tissues. Here the street system is entirely unplanned with all irregularities, and hence the name "Studied Irregularities". The streets are curving, sometimes discontinuous, with variable width and irregular open spaces at some intervals.

To make this system attractive and pleasing with good scenic effects, the streets are planned in conformity with natural surroundings, such as river, lake, hillock, sea-shore etc.

E.g. Carcassonne, France.

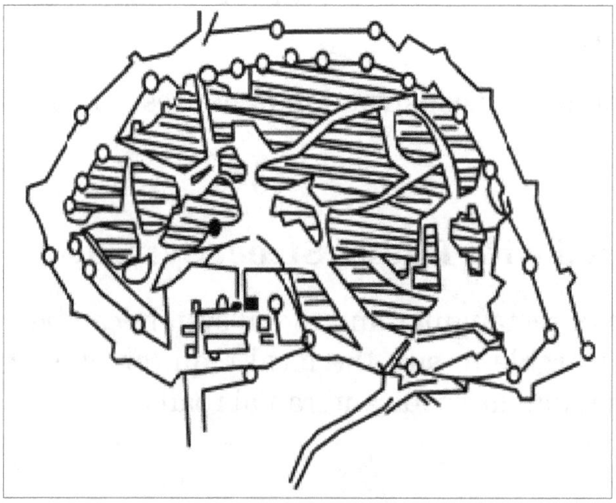

Organic street system.

Irregular Medieval Street System

Many of the old cities of the middle ages have grown in this irregular way without any planning. They have developed immethodically and illogically beyond control.

Combination of Rectangular and Irregular Street Systems

The irregular type is of recent development of the town. When the old system is found difficult to rectify to suit the modern needs, the combination is adopted E.g. Edinburgh, U.K.

4

The Process of Urbanization

The process of a shift in the population from rural areas to urban areas is called urbanization. Some of the concepts studied in relation to urbanization are suburbanization, overurbanization, counterurbanization and peri-urbanisation. All these diverse processes related to urbanization have been carefully analyzed in this chapter.

Urbanization

Urbanization is a process whereby populations move from rural to urban area, enabling cities and towns to grow. It can also be termed as the progressive increase of the number of people living in towns and cities. It is highly influenced by the notion that cities and towns have achieved better economic, political, and social mileages compared to the rural areas.

Urbanization is a population shift from rural to urban areas, "the gradual increase in the proportion of people living in urban areas", and the ways in which each society adapts to the change.

Urbanization is not merely a modern phenomenon, but a rapid and historic transformation of human social roots on a global scale, whereby predominantly rural culture is being rapidly replaced by predominantly urban culture. The first major change in settlement patterns was the accumulation of huntergatherers into villages many thousand years ago. Village culture is characterized by common bloodlines, intimate relationships, and communal behavior whereas urban culture is characterized by distant bloodlines, unfamiliar relations, and competitive behavior. This unprecedented movement of people is forecast to continue and intensify during the next few decades, mushrooming cities to sizes unthinkable only a century ago.

Urbanization is simply defined as the shift from rural to an urban society, which is triggered by social, economic, and political developments. Urban expansion in India will happen at a speed quite unlike anything the country or the world has seen before. It took nearly 40 years (from 1971 to 2008) for India's urban population to rise by nearly 230 million; it will take only half that time to add the next 250 million.

According to a survey by UN State of the World Population report in 2007, by 2030, 40.76% of country's population is expected to reside in urban areas. As per World Bank, India, along with China, Indonesia, Nigeria, and the United States, will lead the world's urban population surge.

Urbanization is pervasive and recent phenomenon. In present global atmosphere, all nations undergo with the challenges of environment, social, transportation, economy in their respective cities. These issues are commonly occurred in developing countries due to the difference of development in cities and villages.

Causes of Urbanization

Industrialization

Industrialization is a trend representing a shift from the old agricultural economics to novel non-agricultural economy, which creates a modernized society. Through industrial revolution, more people have been attracted to move from rural to urban areas on the account of improved employment opportunities. Industrialization has increased employment opportunities by giving people the chance to work in modern sectors in job categories that aids to stir economic developments.

Commercialization

Commerce and trade play a major role in urbanization. The distribution of goods and services and commercial transactions in the modern era has developed modern marketing institutions and exchange methods that have tremendously given rise to the growth of towns and cities. Commercialization and trade comes with the general perception that the towns and cities offer better commercial opportunities and returns compared to the rural areas.

Social Benefits and Services

There are numerous social benefits attributed to life in the cities and towns. Examples include better educational facilities, better living standards, better sanitation and housing, better health care, better recreation facilities, and better social life in general. On this account, more and more people are prompted to migrate into cities and towns to obtain the wide variety of social benefits and services which are unavailable in the rural areas.

Employment Opportunities

In cities and towns, there are ample job opportunities that continually draw people from the rural areas to seek better livelihood. Therefore, the majority of people frequently migrate into urban areas to access well paying jobs as urban areas have countless employment opportunities in all developmental sectors such as public health, education, transport, sports and recreation, industries, and business enterprises. Services and industries generate and increase higher value-added jobs, and this leads to more employment opportunities.

Modernization and Changes in the Mode of Living

Modernization plays a very important role in the process of urbanization. As urban areas become more technology savvy together with highly sophisticated communication, infrastructure, medical facilities, dressing code, enlightenment, liberalization, and other social amenities availability, people believe they can lead a happy life in cities. In urban areas, people also embrace changes in the modes of living namely residential habits, attitudes, dressing, food, and beliefs. As a result, people migrate to cities and the cities grow by absorbing the growing number of people day after day.

Rural Urban Transformation

As localities become more fruitful and prosperous due to the discovery of minerals, resource exploitation, or agricultural activities, cities start emerging as the rural areas transform to urbanism. The increase in productivity leads to economic growth and higher value-added employment opportunities.

This brings about the need to develop better infrastructure, better education institutions, better health facilities, better transportation networks, establishment of banking institutions, better governance, and better housing. As this takes place, rural communities start to adopt the urban culture and ultimately become urban centers that continue to grow as more people move to such locations in search of a better life.

Transportation and Communication

Transportation and Communication is another reason of urbanization, Industries depends on transportation, so, that the raw materials and manufactured goods can be transferred in large volume. The cities are connected with other parts of the country. Communication facilities usually available in urban centers, such as telephones, telegraph, far, computer, wireless and postal services etc. These are the facilities due to which the people move from rural centers to cities. So, in the development of society, transport and communication play a significant role and is a best source of urbanization.

Economic Pull of the City

Cities provide more opportunities of employment and economic facilities than rural areas. So, many people move towards cities for the sake of employment. The businessmen also come to the city from the village to gain better opportunities for employment and economy.

Better Opportunities of Education

In cities, there are better opportunities of education as compare to rural areas. Mostly the best standard schools and colleges are available in the urban areas. Professional colleges and universities such as medical and engineering are situated in urban areas. So, young men and women are attracted to the cities for better and higher education.

Better Recreational Facilities

People migrate to cities for the availability of easy and good recreational facilities cities have move recreational facilities such as parks, play grounds, stadiums, cinemas and clubs. These facilities are not available in rural areas and people tend to move towards urban areas.

Desire for Change in Social Life

Most of the rural people are shifting towards cities to bring change in their social life. The relation between caste and profession is not necessary in cities but the poor people can get better occupational facilities.

Change in the Mode of Living

The changes in the mode of living such as dress, food, attitudes, habits and patterns of residence are also encouraging the people to move towards the urban areas. Village life is traditional and is based on old beliefs, habits and practices. Thus many old things may be changed in urban culture.

Effects of Urbanization

Positive Effects of Urbanization

Urbanization yields several positive effects if it happens within the appropriate limits. Some of the positive implications of urbanization therefore include creation of employment opportunities, technological and infrastructural advancements, improved transportation and communication, quality educational and medical facilities, and improved standards of living. However, extensive urbanization mostly results in adverse effects. Below listed points are few of them.

Housing Problems

Urbanization attracts people to cities and towns which lead to high population increase. With the increase in the number of people living in urban centers, there is continued scarcity of houses. This is due to insufficient expansion space for housing and public utilities, poverty, unemployment, and costly building materials which can only be afforded by few individuals.

Overcrowding

Overcrowding is a situation whereby a huge number of people live in a small space. This form of congestion in urban areas is consistent because of overpopulation and it is an aspect that increases day by day as more people and immigrants move into cities and towns in search of better life. Most people from rural or undeveloped areas always have the urge of migrating into the city that normally leads to congestion of people within a small area.

Unemployment

The problem of joblessness is highest in urban areas and it is even higher among the educated people. It is estimated that more than half of unemployed youths around the globe live in metropolitan cities. And, as much as income in urban areas is high, the costs of living make the incomes to seem horribly low. The increasing relocation of people from rural or developing areas to urban areas is the leading cause of urban unemployment.

Development of Slums

The cost of living in urban areas is very high. When this is combined with random and unexpected growth as well as unemployment, there is the spread of unlawful resident settlements represented by slums and squatters. The growth of slums and squatters in urban areas is even further exacerbated by fast-paced industrialization, lack of developed land for housing, large influx of rural immigrants to the cities in search of better life, and the elevated prices of land beyond the reach of the urban poor.

Water and Sanitation Problems

Because of overpopulation and rapid population increase in most urban centers, it is common to find there are inadequate sewage facilities. Municipalities and local governments are faced with serious resource crisis in the management of sewage facilities. As a result, sanitation becomes poor and sewages flow chaotically, and they are drained into neighboring streams, rivers, lakes, or seas. Eventually, communicable diseases such as typhoid, dysentery, plague, and diarrhea spread very fast leading to suffering and even deaths. Overcrowding also highly contributes to water scarcity as supply falls short of demand.

Poor Health and Spread of Diseases

The social, economic and living conditions in congested urban areas affects access and utilization of public health care services. Slum areas in particular experience poor sanitation and insufficient water supply which generally make slum populations susceptible to communicable diseases. The environmental problems such as urban pollution also cause many health problems namely allergies, asthma, infertility, food poisoning, cancer and even premature deaths.

Traffic Congestion

When more people move to towns and cities, one of the major challenges posed is in the transport system. More people means increased number of vehicles which leads to traffic congestion and vehicular pollution. Many people in urban areas drive to work and this creates a severe traffic problem, especially during the rush hours. Also as the cities grow in dimension, people will move to shop and access other social needs/wants which often cause traffic congestion and blockage.

Urban Crime

Issues of lack of resources, overcrowding, unemployment, poverty, and lack of social services and education habitually leads to many social problems including violence, drug abuse, and crime. Most of the crimes such as murder, rape, kidnapping, riots, assault, theft, robbery, and hijacking are reported to be more prominent in the urban vicinities. Besides, poverty related crimes are the highest in fast-growing urban regions. These acts of urban crime normally upset the peace and tranquility of cities/towns.

Solutions of Urbanization

Building Sustainable and Environmentally Friendly Cities

Governments should pass laws that plan and provide environmentally sound cities and smart growth techniques, considering that people should not reside in unsafe and polluted areas. The objective here is to build sustainable cities that embrace improved environmental conditions and safe habitats for all urban populations. Governments should also encourage sustainable use of urban resources and support an economy based on sustainable environment such as investment in green infrastructure, sustainable industries, recycling and environmental campaigns, pollution management, renewable energy, green public transportation, and water recycling and reclamation.

Provision of Essential Services

Urban stakeholders must ensure all populations within the urban areas have access to adequate essential social services namely education, health, sanitation and clean water, technology, electricity, and food. The objective here is to provide and implement employment opportunities and wealth creation activities so that people can earn a living to pay for the maintenance of the services. Subsidies can also be availed by the government to lower the costs of basic healthcare, basic education, energy, education, public transportation, communication systems and technology.

Creation of More Jobs

To lessen the negative effects of rapid urbanization while at the same time conserving natural ecosystems, private investments should be encouraged so as to utilize natural resources and create more job opportunities. Tourism promotion and the sustainable exploitation of natural resources can create more jobs for the urban populations. Subsidies and grants may as well be provided to foreign and private investment in environmentally friendly development projects that encourage job creation.

Population Control

Key stakeholders in urban areas must provide campaigns and counseling for effective medical health clinics and family planning to help reduce the high rates of population growth. Medical health clinics oriented towards family planning options must be made accessible across the entire urban area with the objective of controlling diseases and population growth.

Suburbanization

Suburbanization describes the general trend of city dwellers to move from the city into residential areas in ever-growing concentric circles away from the city's core. The trend began briefly in the nineteenth century and then exploded after World War II (1939–1945). Suburbs Developed their own shopping and service districts and bred their own distinct lifestyles.

Historians and sociologists have long had difficulty defining the start of suburbanization. Researchers have noted that even before the American Revolution (1775– 1783), land speculators were offering land outside of cities like Boston and New York for development as residential areas. The term "suburban" was in use.

Crowded Urban Centers

Large American cities at that time were heterogeneous. They contained no distinct boundaries for commerce centers, artisan shops, stores, or residences. All existed side by side, often with one building performing several different functions. Indeed, contiguousness was one of the hallmarks of presuburban American cities. Kenneth T. Jackson, author of the classic *Crabgrass Frontier: The Suburbanization of the United States* (1985), has termed these areas "walking cities." Jackson notes that walking cities were very congested. Lots were small, usually no more than twenty feet wide; streets were narrow; buildings were crowded together and tended to grow up rather than out. Any visitor to modern Wall Street—one of lower Manhattan's earliest developed streets—will quickly note the narrow, choked character of the street. People added to the congestion. While early American city population densities would be something less, average European densities ranged from 35,000 to 75,000 people per square mile. Because of the congestion, but also because foot- and horse-travel were the only transportation options for city denizens, inhabitants lived close to their work. Indeed, many lived in apartments above or behind their shops.

City elites tended to congregate near the centers of power, which included governmental buildings and dockside entry points. In the preindustrial days, those were the true seats of commerce. The first areas termed "suburb," in fact, were not pleasant escape destinations for city lead-ers, but rather odious peripheral sites where city law relegated such malodorous businesses as tanneries, slaughterhouses, and soap-making shops. On the periphery, houses of prostitution, vagrant camps, and, later, homes for urban slaves also grew. When the elite did begin moving away from the bustle of commerce centers, it was usually because commercial real estate developers offered them lucrative sums for their homes, or perhaps they wanted to enjoy a country life more fitting a moneyed, genteel society. However, it was important that their homes remained within easy walking or carriage distance of commerce.

Transportation Advances

Transportation advances were the key to modern suburbanization. One of the first was the adoption from Europe of omnibus lines. An omnibus was a large, horse-pulled carriage capable of carrying many passengers. Omnibus lines typically ran to and from city centers, dropping off and taking on paying passengers at different established depots along their line.

After the advent of steam power in the early nineteenth century, it was but a short time before steamer ferries offered easy and practical escape from city congestion. In 1810, using technology perfected by Scottish inventor James Watt, Robert Fulton demonstrated the effectiveness of steam power for river travel. By 1814, steamer ferries connected Manhattan with Brooklyn across the East River. Brooklyn, then isolated and rural, offered a pastoral setting away from the bustling docks and Wall Street for New York's new wealthier class.

Steam-powered railroads, another English invention but one that Americans would exploit to magnificent effect, also helped drive early suburbs. New York City had railroad commuter service by 1832; Boston had it by 1834. Railroads helped transform suburbs into higher-class areas because only the wealthier citizens of a city could afford the rail fare.

Industrial Growth and Immigration

The Industrial Revolution, which began in the United States in the 1830s and continued into the 1880s, had a major impact on the creation of suburbs. Industries, located at first near ocean or river ports, situated themselves near established cities. They drew workers from rural areas, adding to the urban congestion. As more people crowded in, a percentage of people left for the suburbs.

Foreign immigration also fed the urban flight. Revolutions across Europe in 1848, the Irish potato famine of the 1840s, and nationalist unification movements in Italy and Germany in the 1860s all created a class of European émigrés seeking relief in the United States. Immigrants arrived in the United States full of hope, but frequently found those hopes dashed when they discovered it difficult to escape urban areas. Ward heelers working for city municipal "boss systems" or "machines" often met immigrants at their point of entry, gave them a grub-stake of money, and offered them jobs and residences—both in the crowded urban core—in return for perpetual political support. To scared new arrivals, the offers sounded good, but they trapped many in cities, compounding the congestion and grime.

Suburban Growth

As urban cores grew, suburbs grew by a parallel rate. For example, New York City continued to grow throughout the nineteenth century. By the 1890 census, it held more than 2.5 million people. Suburban Brooklyn's population always remained less than half that of New York City's. By the 1890 census, Brooklyn had slightly more than 800,000 people. Philadelphia and Boston experienced similar urban/suburban growth rates.

As Jackson points out, the movement of middle and upper classes to suburbs was "not inevitable." Indeed, such had not necessarily occurred in the great cities of Europe. But Americans carried with them certain emotional, romantic baggage leftover from the early days of the Republic. Thomas Jefferson and other Enlightenment thinkers had vigorously connected republican virtue with agrarian (or at least rural) lifestyles. Reformers during and after the Second Great Awakening (c. 1825–1840) equated rural lifestyles with health. Periodic smallpox, yellow fever, and dysentery outbreaks in crowded cities tended to prove them correct. But the cities could not be erased: by 1870, cities and agricultural areas existed in a state of mutual support, and too many urbanites made their living off the commercial centers.

People with enough money to achieve mobility, however, could enjoy the benefits, both healthful and aesthetic, of the country by moving to the suburbs. In sub-urban homes they would not have to rely on subsistence gardens for food, as rail transportation ensured adequate produce from the countryside. Rather, they could turn their grounds into romantic garden areas, replete with manicured lawns, shrubs, and relaxing fountains.

The Automobile Age

The twentieth century saw the rise of the machine that would cause the explosion of modern suburbia—the automobile. Like so much other technology, autos came to the United States from Europe. But it took American entrepreneur Henry Ford to make them accessible to average Americans. Using the assembly line system to speed and cheapen the production of cars, then enabling Americans to buy them on credit, Ford revolutionized not only the auto industry, but American roadways and city growth as well.

After World War I , when the United States experienced an economic boom that lasted until 1929, modern suburbia began to take shape. The Ford Model T, long the staple of American auto consumers, was quite durable and could, in fact, negotiate dirt roads and ditches, but paved roads became the preference of the driving public. Macadamized roads led into the suburbs, creating a freeway system that fed both metropolitan area sprawl and urban deterioration. As historian Lewis Munford criticized in his book *City in History: Its Origins, Its Transformations, and Its Prospects* (1961), city planners began to subjugate every other consideration to the car. Still, suburban growth was slow. Only 17 percent of the nation's population lived in suburbia in 1920. As with everything else, the Great Depression retarded growth. By 1940 only 20 percent of the population was classified as suburban.

The great suburban explosion, however, came after World War II (1939–1945). In 1945 and 1946, millions of servicemen returned home, all entitled to education benefits under the G.I. Bill and housing benefits under the Veterans Administration (VA) and Federal Housing Administration (FHA). The subsequent issuance of thousands of new bachelors' degrees created a new professional class in the United States, and those veterans who chose vocational education beefed up the service sector. Both classes began having the children that would become the baby boomers, and they needed low-cost housing. Despite the federal backing of home loans, a postwar housing shortage proved an obstacle. The suburbs provided a resolution.

Suburbs, in fact, were an ideal solution to the problem, for transportation was no longer an obstacle. New car sales slowed during the war for obvious reasons: first, men—the primary buyers of cars then—were out of the country by the millions; second, gasoline rationing impeded travel during the war; third, American auto manufacturers produced no new models after the 1942 model year, their plants turned over to military production until the end of the war.

But after Japan surrendered in September 1945, Americans began buying cars once again. They bought a staggering 70,000 cars in the last few months of 1945; that number jumped to more than 2 million in 1946, and more than 5 million in 1949. Many American families, with men drawing army pay and women working in war industry plants, emerged from the war with tidy nest eggs. That gave them plenty for a down payment on a house, and it enabled many to buy their first car or perhaps buy a second car. The advent of two-car families eased any commute for the breadwinner from suburbia: he could take one car while his wife had a second vehicle at home for shuttling children to school and running errands.

Levittowns

Nevertheless, with Americans eager to buy, the housing shortage remained. It took enterprising real estate speculators to break the logjam. None became more famous than builders William and Alfred Levitt. William, a salesman, and Alfred, an architect, wanted to apply mass production techniques to housing. They bought land in Hempstead, Long Island, New York, some thirty miles from New York City, then put together teams of non-union builders who erected simple two-bedroom, one-bathroom homes on sixty-foot by one hundred-foot lots. There the Levitts built 17,000 homes that would ultimately house more than 80,000 people. They integrated parks and swimming pools, and buildings for schools and churches, into their neighborhoods. The first "Levittown" was born, offering mass-produced homes and prepackaged neighborhoods.

When the first Levittown houses went on sale, New Yorkers stood in line to get them. The average price for a two-story, Cape Cod–style home was $7,900. Down payments were about $90; monthly payments averaged $58; that at a time when the average family income was about $3,000 a year. The Levitts followed up their Long Island success with Levittowns in Pennsylvania and New Jersey.

Other speculators and developers followed suit, so that more "Levittowns" sprang up outside other American urban areas. Between 1947 and 1960, the population in suburbs grew by 50 to 100 percent. By 1960, one-third of the nation's population lived in suburban areas. In fact, by 1960, a new phenomenon had occurred—that of "strip cities": one, two, or perhaps more urban areas connected by almost continuous suburbs and suburban shopping areas. The U.S. census counted thirteen of those areas by 1960, obviously including Boston to New York to Washington, D.C., but also Albany to Erie, New York; Cleveland to Pittsburgh; Detroit to Muskegon, Michigan; Toledo to Cincinnati; Atlanta to Raleigh; Miami to Jacksonville, Chicago to Milwaukee, St. Louis to Peoria; Kansas City to Sioux Falls; Dallas–Fort Worth to San Antonio–Houston; San Francisco to San Diego; and Seattle to Eugene.

The Levitts became famous. In 1950, William's likeness appeared on the cover of *Time* magazine. But they also became the subject of criticism from architects, sociologists, and city planners. Some said Levittowns were undermining vital urban sectors. In truth, the automobile had been doing that for three decades. Other critics, like Lewis Mumford, decried the homogeneity—the sameness—of the Cape Codhomes in Levittowns.

But of course, one man's cookie-cutter, prefab house is another man's castle. The people moving into Levittowns were not architectural designers or enlightened romanticists. Rather, they were working people desiring, like their suburban forebears, a chance to get out of the congestion, grime, and, with thousands of cars on the streets, pollution of the cities.

Levittowners tended to love their new homes. They had spacious closets and ample cooking room in the kitchens; the Levitts even supplied television sets with each new home! On weekends, homeowners could putter in their new lawns, planting shrubs and trees. Children could safely bicycle up and down the wide streets of their neighborhoods. Adults, accustomed to the dark, cloistered, often lonely existence of the city, suddenly found themselves sharing backyards with neighbors and actually taking time to visit with them. Neighborhood coffee or bridge clubs sprang up, giving mothers a weekly break from housewifely duties. (Routinely in the 1940s and 1950s, it was the man who made the daily commute from suburbia to the city.) The backyard barbecue became a symbol of suburban freedom, and many a couple struck enduring friendships over the grill. Many suburbanites recall the thrill and pride of moving from a crowded city apartment into their own homes; of having the chance to create a new neighborhood as they had never done before. In that, suburbia reflected the great American optimism of the immediate postwar years.

The Downside of Suburbia

But Levittowns and other suburbs were not all "beds of roses." The togetherness and friendships that suburbs offered could often become double-edged swords. Proximity and the fact that neighborhoods, especially front yards, were open—hence many a suburbanite's activities were open to neighborly review and criticism—gave rise to an intense competition often called "keeping up with the Joneses." The second car (especially a Cadillac, Lincoln, or perhaps a sports model), a boat parked out front, the grass in the front yard, even the barbecue grill could all become competitive status symbols. The comings and goings of teenagers was hard to hide, and sometimes harder to explain. The Tuesday afternoon coffee club was both a time for women to get together and an invitation to engage in intense, often cutting, criticism. If a woman's floor was not clean enough or her cups and saucers not spotless, the news was soon all over the neighborhood. The flight to the suburbs also frequently cut families off from traditional support groups such as parents, grandparents, uncles, and aunts. With those connections harder to maintain, many a young woman found herself home alone with nothing but diapers to wash and floors to clean, and, in fact, doubting her abilities to do that properly.

An element of racism also crept into the suburbs. The rise of suburbia did indicate a "white flight" from the urban core, and critics charged the Levitts and other suburban developers with racism. Levittowns and other neighborhoods had restrictive covenants barring African Americans from buying homes there. The covenants also extended to renting. William Levitt, a Jew who had seen what horrors racial prejudice had wrought in Nazi Germany, chafed under the accusations, and he countered that he was only selling the type of neighborhoods that buying suburbanites wanted. In that Levitt had hit the root of the sociological phenomenon of suburbs: most homebuyers were white, and they wanted to settle in homogenous neighborhoods.

Civil rights advocates also charged in the early 1970s that largely white suburban schools were in violation of the desegregation mandate the Supreme Court handed down in the 1954 *Brown v. Topeka Board of Education* case. In *Milliken v. Bradley,* a 1974 case that attempted to desegregate urban and suburban Detroit schools, the Court maintained that no constitutional violation existed, as the suburban schools were not created with overt intent to segregate. The Detroit area school system remained the same.

Urban businessmen also griped about the loss of business to the suburbs. All suburban areas, given their distance from cities, generated their own support services and businesses—grocery stores, five-and-dime stores, gas stations, restaurants, laundries, doctors' offices, movie houses. Businesses left in downtown areas not only suffered from an undermining of their customer pool, but the customers who were left tended to be of a lower class with less disposable income.

Simultaneously, by 1960, New York, Chicago, and other large cities had ceased to grow; soon their population would start to decline as more people fled to the suburbs. Consequently, the urban tax base also dwindled, cutting into revenues for capital improvements. Inner cities started a period of decline and decay. Tenements remained overcrowded; gangs began offering inner-city youth some type of affiliation and made robbery the order of the day; murders increased as police became over-worked and were unable to maintain an authoritarian presence; graffiti defaced public transportation and buildings. By the late 1960s and 1970s, inner cities had become synonymous with—if not stereotypical of—crime, violence, and filth. Public-housing projects (often notoriously known simply as "the projects") sprang up in cities as alternatives to run-down tenements, but they often quickly deteriorated into new centers of crime.

Not until the 1990s did cities effectively get control over, and in fact revitalize, run-down urban areas. Cities as vastly different in size and culture as Boston and Oklahoma City funded projects to clean up downtown areas, attract businesses back to the city, and pull customers back into the city by cleaning up streets, lighting them, ridding them of crime, and building public attractions like parks, river-walks, and sports facilities. By 2000, Mayor Rudolph Giuliani of New York was being praised for drastically lowering the crime and homicide rates there.

While the flight to suburbia has by no means abated, an interesting phenomenon occurred in the early 1990s. Researchers recognized that some professionals were in fact moving out of the suburbs and back into cities. They could spend less time commuting; commutes in some areas, including mileage and traffic tie-ups, might eat up three hours a day, seriously eroding sleep and family time.

Nevertheless, suburban areas continue to grow. The new millennium, however, has seen downtown urban areas strive to create a more beneficially symbiotic relation-ship with suburbs.

Topics within Suburbanization

The Cities Left Behind

As families left the cities for the suburbs, cities started to experience changes. Downtown businesses started to shift towards the suburbs where the customers were. As businesses left the cities, crime and urban decay became problems. Many towns tried to reinvent themselves as entertainment centers. They demolished large sections in urban renewal projects to build downtown malls to lure people back. In "Magic Lands: Western Cityscapes and American Culture after 1940," John Findlay looks at four areas: Disneyland, Sun City Retirement Community, the Stanford Industrial Park and the 1962 World Fair at Seattle. These four "magic lands," sought to attract middle class families, tourists and retirees to help change the surrounding area. In the process Findlay believes they served as a model for other developments throughout the country. However, even the suburbs have witnessed declines. As newer housing and commercial developments were built the older suburban areas began to decline just as cities did earlier.

Isenber's Downtown America disagrees with the idea that cities were on the decline during this time. Instead she holds that as suburbanization was on the rise that cities were constantly reinventing themselves. In this way changes like the demolition of large buildings to make way for smaller buildings or even parking lost were actually tactical decisions made by the appraisal industry which was responding to falling property values by creating spaces that would generate more income.

Deindustrialization

Deindustrialization swept across much of the country in the late 1970s and 80s, as manufacturing jobs were shipped overseas. *Corporate Wasteland: The Landscape and Memory of Deindustrialization* uses a unique approach of oral history and photography, combined with historical analysis to examine the local stories behind deindustrialization. *Beyond the Ruins: The Meanings of Deindustrialization* takes a different approach, with thirteen essays combining to present the unique historical process of deindustrialization in different places and times.

The African American and Immigrant Experience

Andrew Wiese's *Places of Their Own* (2004) examines the driving forces behind increasing African Americans movement to the suburbs during the twentieth century and the obstacles they faced while doing so. According to Wiese, this study is of great importance because "by virtue of numbers alone, black suburbanization is one of the most important demographic movements in the twentieth century United States." It established the suburbs as an "indispensable context for the study of African American life." In addition, the movement signaled an important historical change in American society. The "suburban boom heralded the emergence of a new black middle class, larger and more economically secure than any black elite in the past." It also revealed a change in long-standing patterns of racial inequality: "a loosening of the shackle between race and class and a breaching of residential barriers" that had prevented African Americans for generations "from the most economically vibrant localities and confined them to areas where locational disadvantages reinforced racial inequality." Wiese makes the case that, as it did with the white population, suburbanization played a significant part "in the making of "African American identities in the twentieth century." Wiese defines the suburb as much by its extra-urban but non-rural location as by its composition; he does this in part to counter the existing literature which had normalized the white, middle-class suburban experience. However, this does mean that some housing forms and areas which do not automatically read as suburban, such as apartment buildings, are included in his work.

The suburbanization trend represented a change from the early part of the 20th Century when there was massive migration of blacks from the South to northern cities. This shift in population played a role in the residential movement of whites to the suburbs. Cities came to be viewed as dangerous and crime ridden. Many middle class families thought they would be safer moving to the edge of the city. This, according to Jackson in Crabgrass Frontier, was the opposite of the pattern in many other countries in the world where slums mostly existed outside of the city. In Places of Their Own, Wiese examined the movement to show how blacks equated land ownership to freedom. Transportation systems helped many blacks live in the suburbs and commute to work in the cities. However, the move to the suburbs also meant moving into segregated communities where interaction with other groups was minimal.

Charles Lamb's Housing Segregation in Suburban America since 1960 examines national fair housing policy from 1960 through 2000 in the context of the American presidency and the country's segregated suburban housing market. Arguing that a principal reason for suburban housing segregation lies in Richard Nixon's 1971 fair housing policy, it traces Nixon's housing legacy through each presidential administration from Gerald Ford to Bill Clinton and as detected in the decisions of Nixon's Federal Court appointees.

Sorting Out the New South City also dealt with patterns in African American suburbanization through the case study of Charlotte, North Carolina. As ideas about what constituted "a good place to live" changed from the 1870s to the present so too did ideas about ideal land use. In this study Hanchett shows the ways in which individual choice, national trends, economic changes, industrialization, political threats, and racism came together to transform Charlotte several times over. Charlotte changed from a salt and pepper mixed race, income, and industry model; to a patchwork pattern which began the processes of dividing races, classes, and businesses; to a sectors pattern which used local impulses, national trends, and federal funds to create rigidly divided neighborhoods which separated black from white, rich from poor, and residential from industrial. In this narrative Northern ideals about suburbanization and national ideas about zoning are not embraced until Populism, African American assertiveness, and growing numbers of working class factory workers challenge elite white power structures. Hanchett also focuses on the role African Americans played in this movement, showing how they moved to particular areas over others as a strategy of directed opportunity - an idea which fits with Wiese's narrative.

The Changing Roles of Women

In Bourgeois Utopias, author Robert Fishman attributes most of this suburban migration to the development of the new Nuclear Family, which he argues featured a more introverted character and sought insulation from the seeming chaos of the burgeoning cities. This lent a domestic character to the process, as the importance of home-life took on new significance. This would particularly impact women; for the domestic sphere was the stock and trade of the feminine gender during the period. The mere act of relocating to a peripheral community had the effect of removing women from the mainstream of social life, which caused them to take a greater role in home life. As mistress of a suburban household, a woman assumed responsibility for the daily operations of that facility and for general oversight of family affairs; thus liberating her husband from these tasks which were popularly conceived as being beneath his social position. Fishman argues that this role did empower women by giving them authority to supervise staff and conduct household transactions, but that these gains were diminished by the social isolation women endured as suburban housewives.

During the post-World War II era and through the 1950s, American women willingly became housewives and mothers, dedicated to making a home for their husbands, maintaining sparkling houses, and bringing up children—all while perfectly coiffed and wearing dresses and high heels. And they liked it, in theory at least. In *Not June Cleaver: Women and Gender in Postwar America, 1945-1960,* Joanne Meyerowitz edits a compendium of essays on women and work and activism, on constructions of womanhood, and on women considered sexual and cultural rebels. She classifies the 15 essays in this volume as revisionist history, placing "the domestic stereotype in historical context and questions both its novelty and pervasiveness in the postwar years. during this era, most American women lived, in one way or more, outside the boundaries of the middle-class suburban home."

The roles of women changed through the early period of suburbanization when many women began to work outside the home: for extra income during the Great Depression; for patriotism during World War II. However, after the war many women found themselves being told that their place was at home raising children and taking care of the house. This image was reinforced by television shows such as "Leave It to Beaver" and "Father Knows Best." Many women who had college degrees and even those who had just been working women or had career aspirations did not feel fulfilled by the roles of being a housewife. Daniel Horowitz in Betty Friedan and the Feminine Mystique, suggests that the political climate of McCarthyism in the 1950's drove many left-wing feminists like Friedan and feminists in general underground. Many of these women would return to lead the feminist movement that reemerged later.

Environmental Backlash

Although we might think that concern over suburban sprawl is a relatively new phenomenon, for decades, environmentalist groups and scientists have expressed reservations about this phenomenon. Written in 2001, Adam Rome's *Bulldozer in the Countryside: suburban sprawl and the rise of American environmentalism* examines the protests against the housing construction craze from 1945-1970. It examines how changing technology allowed builders to construct homes in environmentally sensitive areas, resulting in "frequent flooding, costly soil erosion, and drastic changes in wildlife populations."

Rome discusses how scientists have been concerned over suburban sprawl since the 1940's. Developers also have put houses closer together to increase the number of houses per tract. As the suburbs grow, more and more productive land is lost. As the federal government backed off enforcement of measures passed in the late 1970s that were designed to help the environment, many local governments have stepped in with measures of their own. The local initiatives have included protecting farm land from developers and limiting the size of housing developments. Some cities have also considered limiting the size of commercial developments as a way to slow growth.

There is little doubt that the early environmentalists were awakened in the late 1960's and by the 1970's government agencies like the EPA were working to regulate automobile emissions and passing the Clean Air Act, among others. In Auto Mani: Cars, Consumers and the Environment, the interaction of the automobile and the environment is explored. This work offers a counter narrative to the prevailing myth that Detroit ignored the environment. As the author shows, consumers were also culpable in the process by choosing form over function.

The Culture of the Times

America in the 1950s: the world was not so much a stage as a setpiece for TV, the new national phenomenon. It was a time when how things looked--and how we looked--mattered, a decade of design. With *As Seen on TV: The Visual Culture of Everyday Life in the 1950s*, according to Amazon.com, historian Karal Ann Marling takes readers back to those early days of television, when Ike was in the White House and everybody loved Lucy. The author explains TV's tremendous influence: it allowed Mrs. Eisenhower to give the nation the "Mamie Look," and advertised both Disneyland and the big-business "leisure society" created by the 40-hour workweek. Marling also looks into America's love affair with the automobile ("Drive your Chev-ro-lay through the USA," sang Dinah Shore); the importance of Elvis and Betty Crocker; and Cold War politics, featuring Richard Nixon in the kitchen with Nikita Khrushchev. A nostalgic, informative and sometimes funny view of 1950's American culture.

Urbanization Influencing Suburbanization

With *Magic Lands: Western Cityscapes and American Culture after 1940* John M. Findlay analyzes four "magic lands": Disneyland; the Stanford Industrial Park; Sun City, a retirement community near Phoenix; and the Seattle's World Fair of 1962. These "thematic" places are magical because they provided "spatial coherence," which made western cities more legible to the white, middle-class families, tourists, engineers, or retirees they sought to attract, and "thereby helped to transform the surrounding urban landscape as well as the nation's metropolitan areas." Each, he argues, "was consciously created as a highly controlled, exhaustively planned enclave whose design and symbolic values brought to focus what was most original and important in the amorphous landscape of the urban West." These four magic lands not only served as models for other development throughout the West and the entire United States, but they also "became pivotal fixtures on the western scene by serving to bring a sense of order, community, and refinement to a highly fluid society."

These four "magic lands," sought to attract middle class families, tourists and retirees to help change the surrounding area. In the process Findlay believes they served as a model for other developments throughout the country. Findlay takes issue with people who viewed western cities with contempt because their urban cores were spread out and didn't follow the traditional patterns found on the East Coast. Findlay talks about the magic lands to help show how the western part of the U.S. helped redefine what a city or cityscape could or should be. However, even the suburbs have witnessed declines. As newer housing and commercial developments were built the older suburban areas began to decline just as cities did earlier.

Government's Role

Government policies have played a significant role in the suburbanization process. In the post World War II era the government helped the process along with the building of roads, highways and the interstate highway system. In Magic Lands, Findlay argues that Disneyland and the retirement city of Sun City would not have been successful without the highway system to bring people to them. Society became fluid: One could drive onto a highway and escape to another city very quickly. In the post war boom the housing development of Levittown, New York, was built in response to the huge housing shortage. Levittown represented the first housing development where the suburbs were created in a planned manner. To help returning veterans afford homes in Levittown and other developments, the government offered low-interest loans through the Federal Housing Administration and also through the G.I. Bill. The government in some ways was encouraging the transfer of city dwellers to the suburbs.

The role of new federal programs during the New Deal and in the post WWII era, and the way that they helped to expand suburbanization is a common them throughout scholarship on the topic. The Federal Housing Administration (FHA), the Home Owners Loan Corporation (HOLC), the rise in public housing, investment in roads and highways, and business investment all encouraged suburban growth. Sorting Out the New South City, My Blue Heaven, and Places of Their Own each deal with the ways that these federal programs were used to encourage specific kinds of suburbanization, while blocking others. Specifically they discuss the way in which these federal programs strengthened, and in the case of Charlotte in *Sorting out the New South City,* created race based segregation through discriminatory loan practices.

Working Class Suburbs

For a long time the role of working class suburbs was ignored in favor of dealing with the pristine, middle class suburbs of the post-war era. Both My Blue Heaven and Places of Their Own attempt to fill this hole in the scholarship.

Both works which look at working class suburbs see suburbanization beginning long before the post-WWII boom, instead seeing semi-rural ring developments of owner made homes as suburban, despite their lack of a streamlined or traditionally suburban aesthetic.

Looking at the working class Los Angeles suburb of South Gate in "My Blue Heaven," Nicolaides shows the ways that homes became the primary deciding factor between stability and poverty for working class families before the 1940's. After that time the post-war economic boom created more stability, but did nothing to diminish the feeling that ones home was their most important investment and the thing keeping their family from loosing their economic gains. Because of this residents came to vehemently defend their homes against outside threats, particularly the Civil Rights Movement and African American encroachments into their neighborhood. Nicolaides sees this as the roots of working class conservatism and the Silent Majority.

Weise depicts a similar pattern in "Places of Their Own," but from the African American point of view. Weise shows the ways that black families attempted to expand into the suburbs despite economic, social, and legal restrictions which attempted to hold them in particular geographic areas. Despite these restraints Weise shows how it was individual choice, changing tastes, and rising tides of protest which shaped African American suburbanization across the country.

Economic Impacts

The economic impacts of suburbanization have become very evident since the trend began in the 1950s. Changes in infrastructure, industry, real estate development costs, fiscal policies, and diversity of cities have been easily apparent, as "making it to the suburbs", mainly in order to own a home and escape the chaos of urban centers, have become the goals of many American citizens. These impacts have many benefits as well as side effects and are becoming increasingly important in the planning and revitalization of modern cities.

Impact on Urban Industry

Sprawling Freeway near Toronto, Canada with a suburbanized industry area in the background. Note the far distances from office buildings compared to a downtown. Many office buildings in suburban industry areas are set up on large irrigated campus, versus downtowns that have close buildings and very little greenery.

The days of industry dominating the urban cores of cities are diminishing as population decentralization of urban centers increases. Companies increasingly look to build industrial parks in less populated areas, largely for more modern buildings and ample parking, as well as to appease the popular desire to work in less congested areas. Government economic policies that provide incentives for companies to build new structures and lack of incentives to build on Brownfield land also contribute to the flight of industrial development from major cities to surrounding suburban areas. As suburban industrial development becomes increasingly more profitable, it becomes less financially attractive to build in high-density areas. Another impact of industry leaving the city is the reduction of buffer zones separating metropolitan areas, industrial parks and surrounding suburban residential areas. As this land becomes more economically relevant, the value of such properties very often increases, causing many undeveloped landowners to sell their land.

Consequences on Infrastructure

As America continues to sprawl, the cost of the required water lines, sewer lines, and roads could cost more than $21,000 per residential and non-residential development unit, costing the American government $1.12 trillion over the next twenty five years. Along with the costs of infrastructure, existing infrastructure suffers, as most of the government's money that is dedicated to improving infrastructure goes to paying for the new necessities in areas further out from the urban core. As a result, the government will often forgo maintenance on previously built infrastructure.

Impact on Real Estate Development Costs

For residential properties, suburbanization allows for home prices to decrease, so people can drive until they can find an area in which they can afford to buy a home. However, these homes may lack certain things such as parks and access to public transit. Also, the prices of homes in downtown center usually decrease as well to compete with the inexpensive homes in the suburbs. One of the main benefits of living in the suburbs is that one gets a much larger piece of land than one would in the city. Therefore, as the size of lots increases, the supply of housing is more limited. This is to mean that as city growth patterns increase the population increases leading to suburbanisation which hence leads to the under development of real estate since it is a business.

Fiscal Impact

The fiscal deficit grows as a result of suburbanization, mainly because in less densely populated areas, property taxes tend to be lower. Also, because of the typical spread pattern of suburban housing, the lack of variety of housing types, and the greater distance between homes, real estate development and public service costs increase, which in turn increase the deficit of upper levels of government.

Effect on Urban Diversity

As the trend of suburbanization took hold, many of the people who left the city for the suburbs were white. As a result, there was a rise in black home ownership in central cities. As white households left for the suburbs, housing prices in transition neighborhoods fell, which often lowered the cost of home ownership for black households. This trend was stronger in older and denser cities, especially in the northeast and Midwest, because new construction was generally more difficult. As of the 2010 Census, minorities like African Americans, Asian Americans and Indo-Americans have become an increasing large factor in recent suburbanization. Many suburbs now have since 1990 large minority communities in suburban and commuter cities.

Environmental Impacts

With the growth of suburbanization and the spread of people living outside the city this can cause negative impacts on the environment. Suburbanization has been linked to the increase in vehicle mileage, increase land use, and increase in residential energy consumption. From these factors of suburbanization, it has then caused a degradation of air quality, increase usage of natural resources like water and oil, as well as increased amounts of greenhouse gas. With the increased use of vehicles to commute to and from the work place this causes increased use of oil and gas as well as an increase in emissions. With the increase in emissions from vehicles, this then can cause air pollution and degrades the air quality of an area. Suburbanization is growing which causes an increase in housing development which causes an increase in land consumption and available land. Suburbanization has also been linked to increase in natural resource use like water to meet residents' demands and to maintain suburban lawns. Also, with the increase in technology and consumptions of residents there is an increase in energy consumption by the amount of electricity used by residents.

Factors Effecting Suburbanization

Technological Factors

- Transportation (steam ferry, steam railroad, omnibus, horsecar, electric streetcar, automobile).
- Industrialization (steam power, factory system, moving assembly line).
- Construction (balloon-frame house, machine nails, road paving).
- Improved public utilities for suburban areas (water and sewer systems, electricity).
- Lawn mower.

Social/Cultural Factors

- Class (stratification, separation, exclusivity, conspicuous consumption, upward mobility, envy).
- Privacy (expansion of private realm, separation of domestic sphere from public realm of commerce and politics).
- Family ("closed domesticated nuclear family," clearly defined gender roles, "cult of true womanhood," privacy).
- Home Ownership (economic security, good investment, social status, "real" manhood, upward mobility, moral haven, preserves capitalism, lawn as "moat"; prominent among immigrant groups).
- Leisure (lawn as place for leisure, otherwise unproductive; country clubs; amusement parks at ends of trolley lines; "riding habit").
- Anti-Urban Sentiment (fear, distrust; cities as corrupt and corrupting; aggravated by industrialization).
- Romantic Nature Aesthetic (pastoralism, curving roads, English landscape gardens, "the Picturesque"; came about only when fear of nature subsided).
- Architecture (villa, cottage, bungalow).
- Utopianism, Reformism, Progressivism (Perfectionism in Llewellyn Park, Industrial Home Owners Association Number One and Mount Vernon; moral and social improvement through suburbanization).
- Mobility ("riding habit", car=freedom).
- Bigness ("bigger is better, growth is good;" civic boosterism).
- Faith in Progress.

Economic Factors

- Land speculation.

- Cheap land at periphery.
- Advertising (influences culture).
- Expensive property and transportation served to segregate classes.
- Rising wages in early 20th c.

Political Factors

- Political corruption in transportation permits.
- Taxation.
- Pro-capitalist, anti-communist rhetoric surrounding home ownership.
- Civic boosterism.
- Differential treatment of modes of transportation (transit was taxed while roads were subsidized out of general revenue).
- Influence of special interests (transportation policy).

Legal Factors

- Fee simple ownership of property.
- Home equated with homeowner in Anglo-Saxon tradition: right to kill intruders.
- Setback requirements.
- Rules for incorporation simplified in twentieth century.

Demographic Factors

- Increasing population density in industrial city.
- Immigration from abroad.
- Rural-to-urban migration.

Environmental Factors

- Pollution and disease in industrial city.
- Topography (Westchester County, e.g.).

Overurbanization

When urbanization is at greater level in developed countries in comparison with developing countries, it may be called 'over urbanization'. The fact, that, at comparable levels of urbanization, the developed countries of today had a correspondingly greater proportion of their labour force engaged in manufacturing operations had led to the usage of terms like 'Over-Urbanization', 'Urban-Inflation', 'Hyper-Urbanization' etc; to describe the urbanization - phenomenon in the developing countries of today. Rural-Urban migration caused by the 'Rural-Push', rather than the 'Urban-Pull' leads to 'over urbanization'. It means, when urban population is in excess of, what is the really demanded by the urban sector; urban-inflation occurs. Due to over-urbanization, urban misery and rural poverty exist side by side in the dynamic process of urbanization.'

According to the over-urbanization thesis, the rat e of urbanization in Asia, has been much higher than the pace of industrialization in the countries of this region. In other words, i t was contended that, many of the developing Asian-economies were over urbanized due to a large migration of the population from rural to urban areas, because of increasing pressure of population on agricultural land. Challenging the over - urbanization hypothesis, Prof. N.V. Sovani remarks that, the statement that rural migrants end up in the low productivity sector of urban economy, "describes only a part of the reality; and its suggestive implications are darker than they are in actuality. It is true that migrants to towns are absorbed in low productivity employment is found to be, by and large, more productive than the pre-migration employment of the immigrants.

Prof., N.V. Sovani has found that, "the correlation coefficient between urbanization and industrialization had a much higher " I positive value in the case of developing economies that developed economies". He therefore, concluded that, "the pace of urbanization in the underdeveloped countries is much more closely dependent on the pace of industrialization than in the highly industrialized areas". This flies in the face of the entire over-urbanization thesis, at least in the way it has been formulated up to now.'

The analysis of Kamerschen also revealed that, there was no significant relationship between rural pressure and urbanization. However the sign for the co-efficient of rural pressure (agricultural population per hectare of cultivated land) was positive in all but one case. This indicates that, probably a higher pressure of population on agricultural land tends to encourage the process of urbanization. Finally, the argument that, urbanization hampers economic development of underdeveloped countries through the misallocation of available economic resources was not found to be true.

Causes

Sociologist John Shandra states that arguments about the causes of overurbanization fall into five groups:

1. The rural-push and urban-pull perspective.
2. The economic modernization perspective.
3. The political modernization perspective.
4. The neo-Malthusian perspective.
5. The dependency perspective.

Shandra's analysis of several variables related to each of these categories suggested that all of these arguments have significant evidence except for the economic modernization perspective. Recent scholars believe that a variety of these factors are relevant.

The Rural-push and Urban-pull Perspective

The biggest cause of overurbanization emphasized by scholars is rural-urban migration and the "push" factors associated with it, including "increased population, diminished size of holdings, and absentee landlord exactions." Specifically, lower death rates as a result of demographic transition lead to less available land and fewer opportunities for rural residents. The larger process of urbanization is characterized both by these factors that "push" migrants away from their homes as well as factors that "pull" them towards new areas. Davis and the UNESCO report both discuss that overurbanization is affected by the "push" factors away from rural areas being stronger than the "pull" factors. Pull factors towards urban areas include expansion of economic opportunity and the infrastructure of cities as administrative centers Shandra recognizes the relationship between push and pull factors, arguing that rural conditions, specifically environmental scarcity, cause decreasing income, decreased stability, and increased health risks, leading many to respond by migrating to urban areas. For example, drought in Brazil and Deforestation in the Philippines has made many rural residents' former manner of livelihood impossible, forcing them to move to the nearest city. Because migrants are primarily motivated by factors pushing them out of rural areas rather than factors such as demand for labor pulling them to the city, these rural-urban migrants often find themselves unemployed or quitting "low productive agricultural employment to yet another section marked by low productivity employment, namely handicraft production, retail trading, domestic services in urban areas." A study done by sociologist Glenn Firebaugh showed that agricultural density, a strong indicator of land constraint, and the presence plantation agriculture both have significant effects on overurbanization. These findings were later reversed by sociologist Bruce London, who emphasized that urban migration was not the only potential response to agricultural density.

Sovani argues that there is little evidence for the greater role of "push" factor of increased population in rural areas, as even countries where there is little pressure for land experience this phenomenon, but that instead the opportunity for higher income is responsible for the excessive migration and pressure on cities, as the salary for an unproductive job in an urban area was almost always higher than the salary for unproductive work in a rural area. Graves and Sexton also emphasize that individuals move despite negative factors such as overcrowding, suggesting that individuals still see urban migration as an overall benefit. They argue that if the benefits do indeed outweigh the costs for society as a whole, then the term "overurbanization" is not appropriate to describe the phenomenon. Gugler argues that while the benefits outweigh the costs for an individual migrating to an urban area, greater costs such as resource scarcity and widespread unemployment and poverty are present when this occurs at a larger scale.

Sovani also argues that the definition of overurbanization as developed by scholars in the 1950s and 1960s suggests some sort of limits to population density "beyond which the resulting social situation is abnormal," which he argues need to be defined more clearly. Such unsupportable growth would suggest that the cause of overurbanization is urbanization happening too rapidly for a city's level of economic development. Dyckman would call this the "pre-takeoff period." However, several scholars have questioned the validity of the connection between urbanization and industrialization.

The Economic Modernization Perspective

The economic modernization perspective on the causes of overurbanization is based on modernization theory, which argues that a hierarchical progression exists from pre-modern to modern society. An explanation of overurbanization from this perspective was given by sociologist Jeffrey Kentor, who wrote that under modernization theory, urbanization results from development and industrialization creating jobs and infrastructure. This argument has been criticized by those who do not ascribe to the assumption that there is a linear path of development that all countries follow.

The Political Modernization Perspective

Shandra's take on the political modernization perspective asserts that environmental degradation causes overurbanization, because the destruction of natural resources in rural areas lowers production and increases poverty and health risks. Supporters of the political modernization perspective suggest that a strong civil society supports lower levels of overurbanization. The presences of international non-governmental organizations (INGOs) in rural areas, political protests, and democratic government all have the ability of limiting rural-push factors by limiting factors that lead to resource scarcity. INGOs can reduce overurbanization by stimulating alternative employment outside of agriculture, supporting grassroots movements, and improving rural conditions, such as by providing clean water. Considering the role of political protest, Shandra offers the example of the Chipko movement in India, where local women protested deforestation. Protection of this natural resource "eliminated the causes (i.e., income risk and health effects) that facilitate rural to urban migration by protecting a natural resource base by which rural residents in India depended for their existence." Given these considerations, Shandra argues that repressive regimes that do not respond to the desires of the public are more likely to cause higher rates of urbanization than democratic governments.

The Neo-Malthusian Perspective

The neo-Malthusian perspective is closely related to rural-push and urban-pull factors, but it suggests that the cause behind these factors is population growth, which leads to ecological problems, decreasing agricultural activity, and increased rural poverty. These factors then push rural residents to the city.

The Dependency Perspective

The dependency perspective on the causes of overurbanization is based on dependency theory, which argued that economic and political systems rendered less developed countries dependent on developed countries, which used developing countries for resources, labor, and markets. Proponents of the dependency perspective argue that rural-push and urban-pull factors are not only a result of population growth and resource scarcity, but that these factors, among others, are caused by the exploitation of developed countries and the capitalist principles they operate under. This is to say that "a comprehensive understanding of Third World urbanization cannot focus solely on intra-national, rural-push, urban-pull explanations but must explicitly incorporate the impact of international capitalist forces." This holds that the negative rural-push factors are a result of the manipulation of developed countries. Michael Kentor found that dependence on foreign investment had a lagged effect on urbanization, meaning that urbanization rates increased a few years after foreign companies began profiting in developing countries. Jeffrey Timberlake and Michael Kentor found in their analysis of 69 less developed countries that there was a significant relationship between dependence, as measured by level of foreign investment, and overurbanization.

Additionally, a study done by Bruce London found that factors related to dependency were not only connected to rapid urbanization, but also the negative aspects of urbanization such as urban inequality.

Effects

Economic

Davis and Golden did not see overurbanization as an inherently negative phenomenon, but as a statistical fact that would likely correct itself, as "urbanization will fall off sharply or industrialization will gain a new impetus." Expanding on the latter, they suggest that overurbanization could spur industrial growth, modernization of agriculture, and social change. Even in the case of overurbanization, some of the positive effects of urbanization could be present in regards to economic growth, such as the development of more efficient economics due to scale, technological developments, diversity of both products and occupations, as well as "the greater opportunity of occupational and social mobility and greater readiness to adapt." For example, they argue that industrialization supports greater efficiency of agriculture through technology, an advantage for the productivity of rural farmers as well as urban consumers. However, Firebaugh argues that great efficiency is often a result of an increasingly capital-intensive system, which creates inequality between large and small landowners, such as in the Latin American latifundia system. Furthermore, Timberlake and Kentor found in their analysis of economic growth and overurbanization that countries that experienced increases in levels of overurbanization experienced less economic growth. Economic opportunities are lacking due to "saturated urban labor markets" that exclude much of both the rural and urban populations truncated opportunity structures in rural areas. Furthermore, high infrastructural costs stymie growth.

Social

The UNESCO report emphasized the negative effects of overurbanization, detailing "low levels of living" as "inadequate housing, the almost complete absence of mass sanitary facilities, the presence of filth, squalor, repugnant odours, disease and high mortality" and "large urban groups who have little or no access to educational facilities." Several scholars have agreed that overurbanization puts a strain on the wellbeing of urban residents due to the lack of adequate public services.

Davis and Golden also argue that greater density of dissatisfied impoverished masses could improve conditions to the extent that it provokes the government to enact change to avoid revolution. Dyckman agreed that overurbanization lends itself to the potential for revolution, though he saw this as a potentially destabilizing factor as the conditions would lead to social dissatisfaction and seizure of control by revolutionary leaders. He saw informal squatter settlements as breeding ground for revolutionary activity.

Employment

Despite arguing the potential for economic growth, the UNESCO report also states that overurbanization prevents urban areas and countries from utilizing their "potential human and physical resources" due to unemployment, under-employment, and misemployment. The idea that rural-push factors are stronger than urban-pull factors in cases of overurbanization suggests that it is population pressure in rural areas rather than the pull of urban jobs that leads to rural-urban migration. Migrants often end up unemployed, as overall urbanization rates rise faster than industrialization and the expansion of the urban job base.

In addition to high levels of unemployment, overurbanization is characterized by underemployment and misemployment. Underemployment is defined as the "underutilization of labor," or when available laborers are not working at their full capacity due to seasonal variation in production or excess employment of laborers for the amount of work that needs to be done. Misemployment is defined as unproductive labor, meaning that efforts are considered to "contribute little to social welfare," such as the full-time labor of begging.

While these phenomena are all caused by excessive rates of migration to cities, it is notable that unemployment and underemployment are also problems in rural areas as well. Often unemployment in rural areas is what pushes residents to the city, where better economic opportunities are expected.

Proposed Solutions

A UNESCO report that discussed overurbanization in Asia suggested initial proposals that addressed rural-push factors such as lack of economic opportunity and low productivity by improving agricultural technology and supporting rural industries. Furthermore, rural misery could be reduced by bringing industrialization into rural areas to increase employment and wages and to support the development of infrastructure that creates a "more desirable community environment." The UNESCO report also considers the role of governments in committing to providing adequate housing as well as regional planning that takes into consideration social concerns. However, these considerations, among others that propose dealing with unemployment, have been criticized as "skirting the issue by addressing ‹symptoms' of overurbanization" rather than the root cause.

Later authors also emphasized improving rural conditions to combat overurbanization. Gugler suggested channeling more resources to rural areas and fighting the tendency to neglect rural areas with what economist Michael Lipton deemed "urban bias," the tendency to allocate funds and public works to cities, where the elite and middle classes reside.For example, monetary policies that create artificially low prices for agricultural products harm farmers while creating a surplus for the government. Thus a reallocation of resources to agricultural workers would help shift this system that favors urban elites over the rural poor. Sociologists York Bradshaw and Mark Schafer studied the relationship between INGOs and overurbanization and found that state expenditures towards development were less effective than the role of INGOs. While INGOs were shown statistically to decrease overurbanization, the presence of INGOS did not decrease the effects of foreign capital investment, which is considered one of the root causes of overurbanization by dependency theorists. They and Shandra agree that INGOs can play an important role in decreasing overurbanization by supporting rural communities by promoting both economic and infrastructural development as well as the role of civic society.

Counterurbanization

Counter urbanisation is said to occur when an urban area starts losing its population or when there is decrease in the population of an urban area. In other words when population loss of urban core exceeds the population gain of the ring of agglomeration resulting in loss in overall population then counter urbanisation is said to take place.

It can also be defined as a demographic or social process whereby people move from urban areas to rural areas which is generally due to depreciation or over crowding of inner city. This is the condition developed in cities of developed nations as a result of densification and high population growth due to natural growth and migration. This growth also leads to congestion is city to which people reacts by moving outside the cities. This is seen as an option by those who can afford to travel back to city daily for their work, on the other hand Counter Urbanisation is not seen in case of poor cities as poor cannot afford to travel a large distance daily and also because of the nature of work they do requires them to be in proximity of workplace.

Factors Responsible for Counter Urbanisation

Counter urbanisation is considered to be a negative sign for any urban area. Examples of areas undergoing counter urbanisation are very less as the overall population more or less remains same over the time or in a general case increases over time. One such example includes Newcastle-upon Tyne in Britain which underwent decrease in urban population because of the collapse of industrial activity with the core of city. There are number of factors responsible for counter urbanisation which include economic factors, social factors, increasing problems within city, crime rates, migration etc.

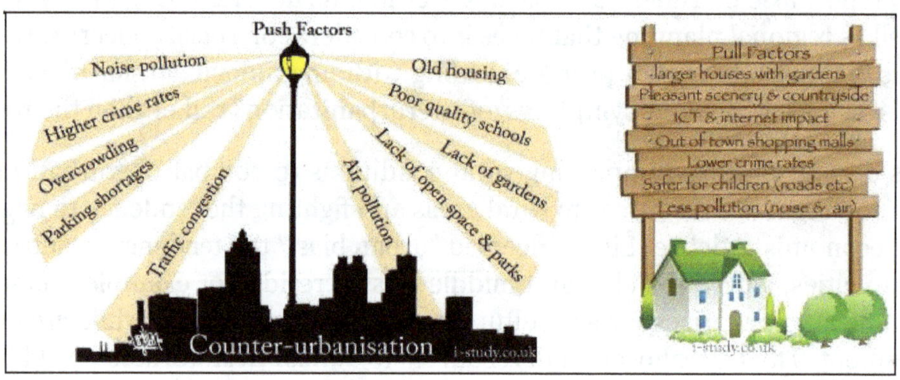

Economic factors: Cities are considered to be engines of economic growth and people migrate to cities in search of better employment opportunities. But when city stops offering the jobs and prospects of better income then the attractiveness of a city decreases. This might happen due to continuous loss of business, change in policies, shift of industries etc.

Migration: This is the most important factor which informs about the counter urbanisation. An overall decrease in population clearly indicates this. The push and pull factors from the affected urban area and the surrounding areas play a large role in triggering the migration in search of better opportunities.

Policies: Policies framed by government can have a large-scale impact over both urban and rural areas. However the impact and the need to regulate activities is more in urban areas as compared to rural areas thus the visible difference is associated with cities. Shifting of polluting activities mainly industries, decentralisation of economic activities, redevelopment projects, population de-densification, increased taxes, inability to provide basic amenities, failure to control crimes, increase in land and housing price etc are responsible for loss of urban population. Government plays a large role in managing and providing the facilities and services which people require on a regular basis.

Technology: Evolution of technology provided a more control over life for mankind, in many cases the need to travel decreased and thus the need to live within an urban area or to visit urban area also reduced in some cases. The rapidly growing concept of work from home is good enough for some to stay and work in rural area than to migrate to cities. Similarly mobile phones, better education and health services in rural areas decreases the need for migration to cities.

Living conditions: Living conditions in an area have a direct impact on the residents. Everyone wish to live at a place with good quality of life and better living conditions. If a city fails to maintain its basic infrastructure or faces epidemic or some other issue affecting masses than it's a warning sign and strong factor of out-migration. Industrial areas are known for poor environmental conditions which have negative impact on health, similar is the case for areas located near mines.

City congestion caused by urbanisation.

There can be a number of other reasons for counter urbanisation which can be people specific, due to deteriorating condition of city, natural calamity, lack of attractiveness or lack of economic opportunities in the urban area.

Possible reason for population decline can be out-migration, decline in birth rate or increase in death rate. People might move to other areas which offer better quality of life and a better price. Cities which neglect spending on social services and public welfare witness counter urbanisation and it needs to be dealt with care. Many of the cities provide incentives in form of tax rebates, relaxations, reduced property tax or development charges to attract people to tackle counter urbanisation.

Effect and Consequences of Counter Urbanisation

Any place needs a threshold population for the activities to continue. A small change in population does not have adverse affect on this threshold population but a large or sudden movement of people or economic activities outside the city might result in collapse of the city. Once the number of people supporting the whole financial system decreases the burden of expenditure to maintain current standard or even to provide the a particular service falls on the remaining population. per capita expenditure of residents increases if they wish to avail that facility or service. This additional burden or the deterioration of the facilities will ultimately force the people who were initially unwilling to leave the city to move to different place. If immediate actions are not taken by the concerned agency than prolonged counter urbanisation results in decay and ultimately collapse or the urban area.

Peri-urbanization

The term peri-urbanization refers to a process, often a highly dynamic one, in which rural areas located on the outskirts of established cities become more urban in character. This transformation occurs in physical, economic, and social terms, and often in piecemeal fashion. Peri-urban development usually involves rapid social change, as small agricultural communities are forced to adjust to an urban or industrial way of life in a very short time. High levels of in-migration are an important driver of social change. Rapid environmental deterioration and infrastructure backlogs are usually another characteristic of the peri-urban landscape. Typically, peri-urbanization is stimulated by an infusion of new investment, generally from outside, including foreign direct investment (FDI).

In spatial terms, Rakodi defines the peri-urban area as:

"the transition zone between fully urbanized land in cities and areas in predominantly agricultural use. It is characterized by mixed land uses and indeterminate inner and outer boundaries, and typically is split between a number of administrative areas."

The peri-urban zone begins just beyond the contiguous built-up urban area and sometimes extends as far as 150 kilometer (km) from the core city, or as in the Chinese case, as far as 300 km. The land that can be characterized as peri-urban shifts over time as cities, and the transition zone itself, expands outward. What frequently results is a constantly changing mosaic of both traditional and modern land use. Peri-urbanization does not necessarily result in an end state that resembles conventional urban or suburban communities. Because so much land is involved, and effective land use guidance systems are virtually non-existent in many countries, it appears that a semi-equilibrium that is neither totally urban nor suburban will result in many cases.

Key characteristics of the peri-urbanization process, particularly in developing countries, include:

- Changing economic structure, encompassing a shift from an agriculturally based to a manufacturing dominated economy.
- Changing employment structure, shifting from agriculture to manufacturing.
- Rapid population growth and urbanization, a phenomenon often not captured in official data because the populations of peri-urban regions tend to be significantly under-counted—in many countries, in-migrants do not officially register as local residents. Many peri-urban areas, furthermore, are still defined as rural, contributing significantly to an undercount of the urban population.
- Changing spatial development patterns and rising land costs.

This highly eclectic and sometimes chaotic pattern of growth produces a monumental public agenda. The pattern of peri-urbanization is often determined initially by the routes of newly built highways, but little of the subsequent growth is properly planned or regulated. A great deal of land speculation is typical, and a boom town atmosphere prevails in which unfettered change outpaces efforts to organize it. Frequently, in developing countries, the requirements of the burgeoning local population for urban infrastructure, housing, medical services, and education are not adequately met. Environmental degradation proceeds rapidly. Industrial enterprises often have to resort to their own private infrastructure to guarantee adequate water, power, and other needs. In those peri-urban areas where in-migration from distant locales is significant, and the ratio of newcomers to long-term residents shifts radically, community building can be a major challenge. Where expectations of employment attract more job seekers than can be accommodated, or where the local poor are left behind, there also emerges in extended urban regions a need to refocus poverty prevention and alleviation efforts to the peri-urban zone.

Drivers

The trend towards the dispersal of population and employment to the peripheries of metropolitan cities is becoming a world-wide phenomenon, but the drivers tend to differ. Large-scale investment, especially in manufacturing, is usually the trigger that sets off peri-urbanization. Often foreign direct investment is the driver, but in some cases, such as China, domestic investment is more significant. Peri-urban regions are attractive to foreign and domestic investment for two main reasons. First, peri-urban regions offer large, relatively inexpensive, land plots and less hindered freight transportation, especially by truck, in support of just-in-time production processes. The need for large land plots is reinforced by investor and government preferences to group manufacturing firms in industrial estates. Industrial estates, especially high quality estates, lessen negative environmental impacts and facilitate government monitoring. These estates also provide locators with reliable infrastructure, spatial clustering of suppliers, often one-window approval services, and an intermediary, or buffer, in dealing with government officials and service providers. At the same time, peri-urban locations enable relatively easy access, usually within a 2.5 hour drive, to a major city that offers advanced producer and personal services, and access to major government decision-makers.

A second driver of peri-urbanization is public policy explicitly supporting dispersal of manufacturing away from core, and even suburban, areas. Generally the underlying rationale for these policies is to ease truck traffic, pollution, and reduce the risk of largescale industrial accidents from manufacturing activities, in the core city. Given the large-scale capital spending involved, and thus their political sensitivity, public policies in support of peri-urbanization are often justified in terms of regional development objectives, i.e., dispersal of employment opportunities, and improvements to the quality of urban life associated with de-industrialization of core cities.

Public investment support for peri-urbanization usually includes the provision of largescale infrastructure, such as ports, highways, rail links, telecommunication facilities, water reservoirs, container handling facilities, airports, and sometimes, publicly-owned industrial estates. These infrastructure investments, usually delivered by national governments, either through line agencies or state enterprises, are often funded through international borrowing. Increasingly, public authorities attempt to attract private investors to fund these large projects through mechanisms such as Build-OperateTransfer. Industrial location incentive packages are usually also a component of the public policy package. Such incentives generally take the form of tariff and corporate tax reductions to investors for a specified period of time, and often include an immigration policy component to enable expatriates to work as high level managerial and technical staff in the industries attracted to the peri-urbanizing areas.

The precise nature of peri-urbanization public policies vary among countries and over time within countries. A striking feature common to most peri-urban areas in developing countries, however, is the lack of sufficient investment in social facilities, city building, and environmental infrastructure. For example, about 88% of cumulative public investment to 1999 in Thailand's flagship Eastern Seaboard peri-urban region has been utilized for "production support infrastructure". Frequently high quality regional plans will be developed for peri-urban areas that include proposals for quality community developments, e.g., new towns, but expected private and public sector investment often does not materialize as planned.

Another driver of peri-urbanization is the availability of relatively inexpensive labor, both in situ, in rural areas that are being enveloped by peri-urbanization, and inmigrants, particularly from poor regions in the countries in question, seeking employment opportunities. Migration dynamics can be either from rural to urban or step-wise from smaller towns and cities. There is wide variation in the mix of migrant versus in situ labor employed in peri-urban areas in different countries with very significant implications for public policy and potential local conflict. Because of labor mobility, the importance of local labor availability is less important than the availability of qualified labor at virtually all skill levels within the country in question that is willing to relocate to peri-urban areas. Daily commutes from the urban core are not possible for the vast majority of labor, especially production workers, because of the long travel times. When labor migration to these regions exceeds employment opportunities, as in the case of the extended Manila region, it frequently leads to hyper-urbanization.

Residential development can also act as a driver of peri-urbanization—this process is termed suburbanization, or in cases where it jumps further out from the core, exurbanization. Middle and upper-middle class groups seeking more space at an affordable price, may purchase, and live in residences in peri-urban areas even though they do not work in the area. This driver is most important where peri-urbanization is found relatively close to the core city, or where core city personal security concerns are high, such as in Manila, Jakarta, and Johannesburg. Large property developers, often single-handedly shape peri-urban localities by building large integrated residential complexes comprising several gated housing developments catering to a range of incomes, and new commercial centers to service them. These gated communities, situated in peri-urban green fields, are essentially privately financed new towns.

Public policy to relocate slum settlements out of the center city, or to relocate rural communities for large projects, such as dams, is another important driver of periurbanization, e.g., in Chongqing, China. Public housing, or suitable sites and services, are not always provided, prompting settlement in fragile, and frequently dangerous, areas, such as wetlands or steep slopes. Relocation decisions rarely take into consideration the availability of employment opportunities and social services. Thus, where population relocation policies are an important factor in the peri-urbanization process, the informal sector tends to play a much more significant role in the peri-urban economy, and access to social services by migrant populations is usually inadequate.

Urban Renewal

Urban renewal, comprehensive scheme to redress a complex of urban problems, including unsanitary, deficient, or obsolete housing; inadequate transportation, sanitation, and other services and facilities; haphazard land use; traffic congestion; and the sociological correlates of urban decay, such as crime. Early efforts usually focused on housing reform and sanitary and public-health measures, followed by growing emphasis on slum clearance and the relocation of population and industry from congested areas to less-crowded sites, as in the garden-city and new-towns movements in Great Britain. Late 20th-century criticisms of urban sprawl prompted new interest in the efficiencies of urban centralization.

Each country approaches urban renewal according to its means and its political and administrative systems. One of the chief activities of urban renewal is redevelopment, which is achieved through the clearance and rebuilding of structures that are deteriorated or obsolete in themselves or are laid out in an unsatisfactory way. Other aspects of urban renewal involve the reuse of the land for new purposes, rehabilitation of structurally sound buildings that have deteriorated or lost their original functions, and conservation—a protective process designed to maintain the function and quality of an area, for instance, by requiring or assisting adequate maintenance while preventing inappropriate development or uncharacteristic changes in the use of land and buildings.

The Importance of Urban Renewal

Economic Growth

SCBD Jakarta was established in 1987. It is one of world-class business district in Jakarta.

The economic Growth means more GFA (Gross Floor Area) to be provided in order to keep up with the flux of population and spike up in businesses. More labours and even more immigrants looking for work. There will be more infrastructure to be developed to ensure all activities, traffic and population increment are well addressed.

Physical Development

Jakarta used to have tram as transportation during dutch colonial. It was developed to bus system then it was converted to Busway.

More GFA can be achieved by building more floors. It means the old typical shop building type - 1 and 2 storeys- must make ways for the high rises and skyscrapers or simply foot-print increment.

Tourism

Tourism is frequently seen as the cornucopia, the classical horn of plenty, bringing employment, higher standards of living and much needed foreign exchange in its wake.

Dempsey and Clarke Quay in Singapore. Before and After renovation. Gentrification as a way to keep historical value and contribute to the Singapore Tourism hence huge revenue for the country. To some extent gentrification causes the place to lost its jinx.

Cultural Heritage (Sociology Concern)

Chinese settlements in Jakarta along glodok area. Past and Present.

Issue of cultural heritage become more popular as scholar and citizens had raised their concerns on balancing urban development between new buildings and tourist attractions. More importantly is to give a sense of belonging to the next generation. One needs to know one's root, one's neighbourhood history, one's cultural history and characteristic of communities in order to value advancement and stay grounded.

The Four Phases of Urban Renewal

Envisioning Change: The Scoping Phase

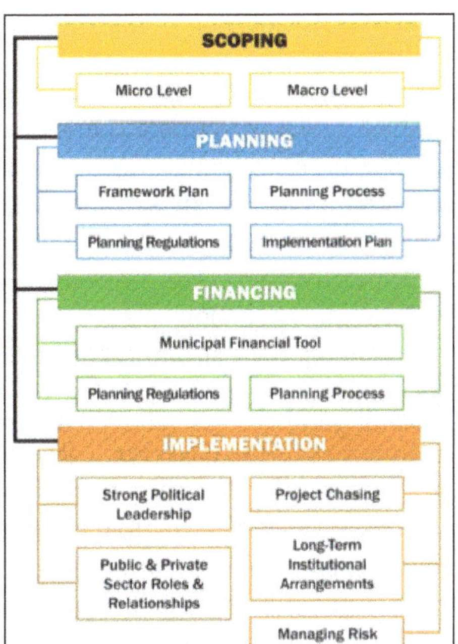

A transformational urban regeneration process, either for the city as a whole or for a specific land parcel—starts with a scoping exercise. Scoping is a process, which provides decision makers with a strategic assessment to identify and promote regeneration. Scoping provides an analytical foundation and engagement process that the city leaders can use to generate choices and to debate and decide on the best course of action. A sound scoping process supplies city leaders with analytical tools to confront issues facing the city that are vital to its future, and to make strategic decisions on the direction they want their city to pursue. Scoping is both forward looking and backward looking. It is forward looking by analyzing what is required for the city and regeneration area to be successful in the future and backward looking in understanding the city's history and the unique "DNA" of the city that are precedent for prudent action and can be capitalized on.

Designing a Web of Actions and Institutions: The Planning Phase

While scoping phase provides an analytical foundation and an overarching rationale and narrative for the regeneration project, the planning framework establishes the long-term vision and context. It is vital to sustaining the regeneration vision through the inevitable changes and unforeseen challenges of market and political cycles. However, it is equally important not to create an overly strict planning structure, which rigidly regulates the outcomes as it discourages private investment. An effective planning framework will balance vision, planning principles and facilitates negotiation amongst the public, private and community sectors.

A sound planning framework couples an inspiring vision with a clear regulatory process. This will provide the private sector with the assurance to invest and take risk, and assures communities that public goals will be achieved and not subordinated purely to the dictates of the market. Using the findings in the scoping section, the planning phase of a regeneration process starts with laying out the foundation on how to prepare for the planning process. The planning phase must detail all the vital elements or assets of the regeneration project, including land, community, and the environmental issues.

Incentivizing Private Funding: The Financing Phase

Whether an urban regeneration initiative is public sector or private sector led affects the types of financing tools that may be accessible to its lead sponsor. Large-scale urban regeneration projects are complex and require immense resources to be planned and implemented properly. Therefore very few cities have the resources to finance all the costs of such large initiatives outright. The partnership with the private sector is not only necessary to share the costs but it is also needed to share the risks and technical capacities. Typically, therefore, cities deploy a combination of internal and external funding sources, policy and regulatory tools, and strategic partnerships with the private sector, among other strategies for financing their urban regeneration vision.

Many factors affect the methods used for financing urban regeneration initiatives. The most important factor is the national legal and institutional context with respect to the amount of control a city has over its fiscal management, specifically its ability to raise and allocate revenue. For example, a city's ability and creditworthiness to access municipal bond markets is a determining issues. Another important factor is the city's capacity to use innovative financing techniques, such as tax increment financing or development rights transfers to support specific real estate and economic development strategies. Based on the case studies in this volume, there are two distinctive groups of instruments to be used for financing urban regeneration: financial tools and regulatory tools. Financial tools involve direct financing assistance for a regeneration project such as various value capture methods (impact fees, special assessments, exactions). Regulatory instruments are essential in leveraging a city's regulatory powers to incentivize private sector participation. An example of this type of tools is tax-based and non-tax-based incentives, zoning, land-use regulations, and development rights transfers.

Evolution from Ideas to Action: The Implementation Phase

Implementation phase entails translating the vision for long-term change into the financial, contractual and institutional relationship between the public and private sectors. This phase includes structuring sound institutions and an organizational structure that is viable and sustainable and can exist through multiple political administrations. It also involves developing sound contracts to translate the vision into a tangible partnership between public and private sectors.

Political leadership may be the most important factor in the implementation phase. As regeneration is a long term, transformative process of change that entails disruption and risk, political leadership is essential to managing the change process so that all stakeholders feel engaged in the process, understand its importance to the future of the city and have genuine outlets to participate. The longitudinal test of political leadership is whether the vision is sustained through changes in political and economic cycles and results not in community division but greater cohesion.

Another major indicator of success in the implementation phase is to clearly map out the life of the project and the expected project cycle and phases as best as possible, accounting for uncertainties. Up next is the phasing and breakdown of a large project into manageable components. The handover process must account for the complex set of interdependencies between the large scale investments and construction projects, such as major infrastructure (utilities, transport, park systems, brownfield cleanup), and the microdelivery of commercial and residential projects that need to establish a "sense of place" and feel complete even though the ultimate build-out of the regeneration project will progress over a longer period. One last important component of implementation arrangement is determining the optimum institutional structure for project delivery and assigning the powers and tools to various entities. This affects the structure of contracts, handover, and sustainability of the project.

Principles for Better Urban Renewal

1. Create "Shared Value"

Urban areas do not belong to a single group or individual but should offer value to many actors. All those who are part of the broader community as a whole – from workers and tourists, children and students, to the underserved and investors – should benefit from urban renewal. "Ultimately, the 'communities' for whom the value is created to share, should be those with long-term interests, not transient stakeholders with a primary focus on value extraction and repatriation," write the authors.

2. Plan With Input From All

Delivering this shared value requires engaging with communities. Planners bringing an intervention into an existing space should share their vision and include people in the planning from day one, or risk it being rejected. Decision-making techniques such as cost-benefit analysis should be explained and employed to also promote "non-financial values," helping communities feel a sense of ownership. The researchers also suggest the creation of a common platform where information about the process and the progress of the project can be shared transparently.

Superkilen Park in the Nørrebro neighborhood of Copenhagen.
The work brings together elements from many of the more than 50 nationalities represented in the area.

3. Build a Long-Term Vision

In any extensive process of urban renewal, the initial goals of the project may change over time. Even so, a long-term vision should be locked in and changes for the sake of short-terms gains resisted, with flexibility growing as the timeline extends further into the future. "A commitment to the public interest and shared value needs an inclusive approach, and future development stages should have the flexibility to be able to adapt to market and social changes," write the authors.

4. Agree on Non-Negotiables

Non-negotiation issues should be clearly understood by all stakeholders. These could include respecting existing lease terms, fixed quotas for affordable housing or protecting open spaces. The rights of renters and leaseholders should be guaranteed and stakeholders agree to a common set of design standards.

5. Agree on a Financial Profile

Studying how the space to be renovated is expected to yield from a financial perspective not only serves to set parameters for the project's development options but is also critical to whether the public's interests will be met. There are many options available to both provide returns on government investment in underserved areas but also safeguard communities from potential negative side effects, like rising taxes, and encourage a handoff to private developers in the future.

6. Establish Clear Development Goals

The planning process should develop and affirm clear objectives, not just desired outcomes. The best goals will be specific and measurable, and anticipate the physical, economic and social results of the project.

High Line Park in Manhattan is built over a disused elevated rail line.

7. Establish Options to Achieve Development Goals

There are often multiple options for achieving the same development goals and they should be compared to one another as well as to baseline scenarios of what might happen without any intervention. The process will create "a much clearer picture of marginal benefits and costs associated with any particular development option," the report explains.

8. Incorporate a Sense of "Localness"

Local characteristics and peculiarities should be captured and incorporated into the new project. These details may come from local standards, services offered in the region, the environment, the climate or other socio-cultural specificities. Finding ways to assimilate a sense of the local into the project will help people identify with it, separate it from other similar projects and generate community acceptance.

9. Evaluate Options With the Goal of Maximizing Net Community Benefits

Cost-benefit analyses are often viewed with skepticism, but the report notes there are well-documented techniques that allow for the inclusion of things communities care about most, like open space, social capital and heritage. Finding ways to incorporate them into a cost-benefit analysis is important for avoiding the scenario where "financial considerations or otherwise vague community aims end up dominating choices between options."

10. Align the Procurement Model With the Planning Vision

Finally, the governance, implementation and contracting trajectories – how the project is actually carried out – should align with the unique vision laid out during the planning phase. Procurement targets should bespoke, rather than using "off-the-shelf" options. This may mean a greater role for the government as a developer in the early stages before handing off to private sector developers later.

"These principles are not locked in place, but are guiding principles to ensure urban renewal benefits the widest community possible," write the authors. "The renewal of strategically important urban sites must have the needs of communities, both social and commercial, at its core."

References

- Andres, Duany. Suburban nation : the rise of sprawl and the decline of the American Dream. Plater-Zyberk, Elizabeth, Speck, Jeff (10th anniversary ed.). New York. ISBN 9780865477506. OCLC 555656296
- Importance-urban-redevelopment-renewal-regeneration-tirtawijaya: linkedin.com, Retrieved 03 May 2020
- Boustan, Leah Platt; Margo, Robert A. (2013). "A Silver Lining to White Flight? White Suburbanization and African-American Homeownership, 1940–1980". Journal of Urban Economics. 78: 71–80. doi:10.1016/j.jue.2013.08.001
- Causes-effects-solutions-urbanization: conserve-energy-future.com, Retrieved 01 March 2020
- Opp, Susan M.; Herberle, Lauren (June 28, 2008). Local Sustainable Urban Development in a Globalized World. Florence, Kentucky: Routledge. ISBN 0754649946

5

Geographical Study of Population

The relationship between the terrain and the composition, migration, distribution and growth of populations is studied under the discipline of population geography. Some of the major topics studied within population geography are demography, mortality rate and population policy. This chapter discusses in detail these aspects of geographical study of population.

Population Geography

Population geography is a branch of human geography that is focused on the scientific study of people, their spatial distributions and density. To study these factors, population geographers examine the increase and decrease in population, peoples' movements over time, general settlement patterns and other subjects such as occupation and how people form the geographic character of a place. Population geography is closely related to demography (the study of population statistics and trends).

Population geography as an independent sub-field of human geography is a comparatively recent phenomenon. In the expression 'population geography', the term 'population' signifies the subject matter and 'geography' refers to the perspective of investigation. Thus, population geography can be interpreted as the study of population in spatial perspective. Etymologically, population geography implies the investigation into human covering of the earth and its various facets with reference to physical and cultural environment.

In the academic world, any discipline is almost invariably defined by its subject matter Johnston. The subject matter of population geography has been a matter of debate ever since Trewartha formally raised the issue in 1953. So is the case with the definition of the sub-discipline.

According to Trewartha, population geography is concerned with the understanding of the regional differences in the earth's covering of people. "Just as area differentiation is the theme of geography in general, so is of population geography, in particular". Population geography is the area analysis of population which implies "a wider range of population attributes than most geographers have ordinarily included" in their analysis. Trewartha proposed a very comprehensive outline of the content of the sub-discipline, which many subsequent geographers seem to have adhered to.

Broadly speaking, the concerns of population geography, according to Trewartha, can be grouped into three categories:

1. A historical (pre-historic and post-historic) account of population;
2. Dynamics of number, size, distribution and growth patterns;
3. Qualities of population and their regional distribution.

Regarding historical account of population, Trewartha suggested that where direct statistical evidence is not available, geographers should adopt indirect methods, and collaborate with anthropologists, demographers and economic historians. In Trewartha's opinion, an analysis of world population patterns, population dynamics in terms of mortality and fertility, area aspect of over and under population, distribution of population by world regions and settlement types and migration of population (both international and inter-regional) form an important part of analysis in population geography. And finally, with regard to qualities of population, he suggested two broad groups – physical qualities (e.g., race, sex, age, health etc.), and socio-economic qualities (e.g., religion, education, occupation, marital status, stages of economic development, customs, habits etc.).

In his book "A Geography of Population: World Patterns", published in 1969, Trewartha arranged these topics in two parts. While the first included a geographical account of population in the past, the second incorporated all the characteristics of population including biological, social, cultural and economic characteristics.

John I. Clarke, who is credited with bringing out the first textbook on the sub-discipline in 1965 (at least after Trewartha had made the case of population geography in 1953), suggested that population geography is mainly concerned with demonstrating how spatial variation in population and its various attributes like composition, migration and growth are related to the spatial variation in the nature of places.

He opines that the main endeavor of population geography is to unravel the complex relationship between the population phenomena, on the one hand, and cultural environment, on the other. His book on Population Geography (1972) and his treatment of the subject matter is in conformity with that of Trewartha, though not as comprehensive as that of the latter.

W. Zelinsky, a contemporary of Clarke, takes a similar view regarding the definition of population geography. He defines the sub-discipline as "a science that deals with the ways in which geographic character of places is formed by and, in turn, reacts upon a set of population phenomena that vary within it through both space and time as they follow their own behavioural laws, interacting one with another, and with numerous non- demographic phenomena".

On the delineation of the field of population geography, Zelinsky suggested that "the list of human characteristics of practical interest in the population geography may be equated with those appearing in the census schedules and vital registration system of the more statistically advanced nations".

Daniel Noin in 1979, in his book "Geographie de la population", while agreeing with the scheme of Trewartha, expressed that distribution of population, components of its growth and characteristics are the main concerns of population geography. More recently, while discussing the methodological problems in population geography, R.J. Proyer suggested that population geography deals with the analysis and explanation of interrelationship between population phenomena and the geographical character of places as they both vary over space and time. According to him, population phenomena include "the dynamics of population distribution, urban/rural location, density and growth (or decline); mortality, fertility and migration; and structural characteristics including age-sex composition, ethnicity, marital status, economic composition, nationality and religion".

Obviously, delineating the precise field of the sub-discipline has been a major problem before the scholars ever since it's beginning. It has been argued that population geographers have spread themselves too thinly over too large a field that they have not been able to establish a niche for themselves in population studies. Scholars have, therefore, suggested that population geography should narrow its focus and concentrate on the components of population change. Woods has made a distinction between broad definition and narrow definition.

The former is described as an elaboration of Trewartha's wide ranging agenda in which certain primacy is given to spatial variation in population, while the latter refers to approach which prefers analysis of population dynamics, namely, fertility, mortality and migration only. Noin's survey in 1984 of the contents of population geography textbooks that appeared during the preceding two decades revealed that broad definition has been most widely used.

Woods (1979) and Jones (1981) proposing the narrow definition have confined the main concern of population geography to the analysis of fertility, mortality and migration at various scales. They claim that the narrow definitions reflect a new process orientation, contrary to the traditional pattern orientation of broad definitions, and are more in line with current trends in geography as a whole. Woods and Rees (1986) propose the term 'spatial demography' in place of 'population geography', which differs from the latter "mainly in terms of the equal emphasis on mortality, fertility and migration as components of population change and distribution its use of the statistical demographic methods and its multi disciplinary approach" (quoted in Heenan). However, as pointed out by Heenan (1988), the distinction seems to be one of semantics rather than one based on critical or substantial epistemological or methodological differences.

From the above, it is, however, clear that the main difference of opinion is on the main thrust in the sub-discipline and not on approach and methodology per se. Woods himself says that "the role of population geography is to provide the spatial perspective in population studies", and that "population geography should be what geographers active in teaching and research do" (quoted in Heenan). To quote Heenan, "if that is so, then in view of apparently increasing erosion of disciplinary boundaries among subjects of common interest in population studies, a case might be made in favour of a definition encompassing any work in which the perspective is principally and explicitly a spatial one – in other words, such a definition would refer to a kind of approach and supporting methodology, rather than to a more or less exclusive disciplinary orientation".

As Woods himself admitted, the two – broad and narrow definitions – are not mutually exclusive, rather they represent differences of emphasis. They are complementary to each other, and taken together, provide the full diversity of works undertaken by researchers in the field of population geography.

Rightly remarked by Clarke that one cannot possibly do justice to all the aspects of population that appear in the census schedules or vital registration system (as suggested by Zelinsky) and that some will receive more treatment than others, partly because they are more central to the theme of population geography and partly because they have attracted closer attention from geographers.

To conclude, the main concern of population geography revolves round the following three aspects of human population:

1. Size and distribution, including the rural-urban distribution of population.
2. Population dynamics – past and present trends in growth and its spatial manifestation; components of population change, viz., fertility, mortality and migration.
3. Population composition and structure. They include a set of demographic characteristics (such as age-sex structure, marital status and average age at marriage etc.), social characteristics (such as caste, racial/ethnic, religious and linguistic composition of population; literacy and levels of educational attainment etc.), and economic characteristics (such as workforce participation rate and workforce structure etc.).

Human Population

Human population refers to the number of people living in a particular area, from a village to the world as a whole. A secondary meaning of population is the inhabitants themselves, but in most uses population means numbers.

No one knows the population of the earliest humans, but there may have been only a few tens of thousands of individuals when the species Homo sapiens first emerged 200,000 years ago. Today more than 6 billion human beings inhabit the earth. Three-fifths of them live in one continent, Asia, with the rest occupying every continent except Antarctica.

The overwhelming bulk of human population growth has occurred since the Industrial Revolution began, more than half since 1950. All but a small percentage of the roughly 80 million people added to world population each year live in the world's developing countries, which are home to 80 percent of humanity and more than 95 percent of world population growth.

The recent growth of the human population has resulted in intense damage to the biosphere, representing a global environmental crisis. The degradation has occurred on a scale and intensity that is comparable to the enormous effects of such geological processes as glaciation. The impact of the human population on any one region, as on the biosphere as a whole, is a function of two interacting factors:

1. The actual number of people.
2. Their per-capita environmental impact, which largely depends on the degree of industrialization of the society and on the lifestyles of individuals.

In general, more damage is done to Earth to support a person living a highly industrialized lifestyle than to support one living a pretechnical-agricultural or hunter-gatherer lifestyle. However, a direct correlation between industrialized comfort and birth-rate is often observed: the more well-to-do a population is (e.g., that of Europe or the United states), the lower its birthrate tends to be.

Size of the Human Population

The human species, Homo sapiens, is by far the most abundant large animal on Earth. Our world population is growing at about 1.5% annually, that is, at about 80 million people per year. If the percentage of annual growth remains constant, the number of people added yearly increases: 1.5% of 100 is 1.5, but 1.5% of 200 is 3. Thus, 80 million new people per year is an approximate figure valid only for a short space of years. Also, global birth rates vary from year to year. If the current rate of growth is maintained, the size of the human population will double in less than 50 years, at which time there will be more than 12 billion people on Earth.

No other large animal is known to have ever achieved such an enormous abundance. Prior to over-hunting during the nineteenth century, the American bison (Bison bison) numbered 60–80 million animals and may have been the world's most populous wild large animal. The most abundant large animals in the wild now are the white-tailed deer (Odocoileus virginia-nus) of the Americas, with 40–60 million individuals; and the crabeater seal (Lobodon carcinophagus) of Antarctica, with 15–30 million. These species have populations less than 1% of that of human beings at this time.

Large animals domesticated by human beings have also become enormously abundant, and these companion species must be supported by the biosphere in concert with their human patrons. Therefore, they must be considered an important component of the environmental impact of the human population. These animals include about 1.7 billion sheep and goats, 1.3 billion cows, and 0.3 billion horses, camels, and water buffalo. Human beings are also accompanied by enormous populations of smaller domesticated animals, including 10–11 billion chickens and other fowl.

Carrying Capacity and Growth of the Human Population

A population of organisms changes in response to the balance of the rates at which new individuals are added by births and immigration and the rate at which they are lost by deaths and emigration. Zero population growth occurs when the growth and loss parameters are balanced. These demographic relationships hold for all species, including ours.

The history of Homo sapiens extends to somewhat more than one million years. For almost all of that time relatively small populations of human beings were engaged in subsistence lifestyles that involved hunting wild animals and gathering wild edible plants. The global population of human beings during those times may have been as large as a million or so individuals. However, colonization of new parts of the world (e.g., Asia, Europe, Australia, Polynesia, the Americas) and occasional discoveries of new tools and weapons allowed prehistoric human beings to grow in numbers and become more effective at gathering food. This in turn allowed the population to increase.

About 10,000 years ago, the first significant developments of primitive agriculture began to occur. These included the domestication of a few plant and animal species to achieve greater yields of food for human beings. The development of these early agricultural technologies and their associated sociocultural systems allowed enormous increases in environmental carrying capacity for human beings and their domesticated species, so that steady population growth could occur. Even primitive agricultural systems could support many more people than could a subsistence lifestyle based on the hunting and gathering of wild animals and plants.

Further enhancements of Earth's carrying capacity for the human enterprise were achieved through other technological discoveries. For example, the discovery of metals and their alloy—first copper and bronze, later iron and steel—allowed the development of superior tools and weapons. Similarly, the invention of the wheel and of ships made possible the easy transportation of large quantities of valuable commodities from regions of surplus to those of deficit. At the same time, increased yields in agriculture were achieved through a series of advances in breeding of domesticated plants and animals and in farming techniques. The evolution of human sociocultural systems has thus involved a long series of discoveries and innovations that increased the effective carrying capacity of the environment, permitting growth of the population. As a result of this process, there were about 300 million people alive in AD 0, and about 500 million in 1650, at which time the rate of population growth increased significantly. This trend has been maintained to the present. The recent explosive growth of the human population has several causes. Especially important has been the discovery of more effective medical and sanitary technologies, which have greatly decreased death rates (especially infant and child death rates) in most human populations. There have also been enormous advances in the technologies that allow effective extraction of resources, manufacturing, agriculture, transportation, and communications, all of which have allowed further increases in the carrying capacity of the environment.

As a result of these relatively recent developments, the global population of human beings increased from about 500 million in 1650 to over one billion in 1850, two billion in 1930, four billion in 1975, and five billion in 1987. In 2005, the human population was approximately 6.56 billion individuals.

More locally, there have been even greater increases in the rate of growth of some human populations. In recent decades some countries have achieved population growth rates of 4% per year, which if maintained would double the population in only 18 years. One third of all the world's births occur in India and China, the two most populous countries in the world (about 1.049 billion and 1.28 billion persons, respectively).

These sorts of population growth rates place enormous pressure on ecosystems. For example, the human population of central Sudan was 2.9 million in 1917, but it was 18.4 million in 1977, an increase of 6.4 times. During that same period the population of domestic cattle increased by a factor of 20 (to 16 million), camels by 16 times (to 3.7 million), sheep by 12.5 times (to 16 million), and goats by 8.5 times (to 10.4 million). Substantial degradation of the carrying capacity of dry lands in this region of Africa has been caused by these increases in the populations of human beings and their large-mammal symbionts, and there have been other ecological damages as well (e.g., destruction of trees and shrubs for cooking fuel).

Another example of the phenomenon of rapid population growth is the number of people in the province of Rondonia in Amazonian Brazil. This population increased twelvefold between 1970 and 1988, mostly through immigration, while the population of cattle increased by 30 times. These population increases were accompanied by intense ecological damage, as the natural rainforests were "developed" to sustain human beings and their activities. (The areas in question are not, for the most part, "developed" in the sense of being transferred from their wild state to a sustainable agricultural state, but in the sense of being stripped and degraded for short-term profit).

Future Human Population

The growth rate of the global human population achieved a maximum during the late 1960s, when it was 2.1% per year. If sustained, this rate was capable of doubling the population in only 33 years. This rate of increase slowed somewhat to 1.5% per year in the 1999, equivalent to a doubling time of 47 years, and in 2002 had slipped to about 1.3%. Even at today's comparatively modest growth rates, the human population increases by about 80 million people each year.

Reasonable predictions can be made of future increases of the human population. The predictions are based on assumptions about the factors influencing changes in the size of populations, for example in the rates of fecundity, mortality, and other demographic variables. Of course, it is not possible to accurately predict these dynamics because unanticipated changes, or "surprises," may occur. For example, a global war could have an enormous influence on human demographics, as could the emergence of new diseases. AIDS is an example of the latter effect, because this lethal viral disease was unknown prior to the early 1980s.

As a result of these uncertainties, it is not possible to accurately forecast the future abundance of human beings. However, reasonable assumptions about demographic parameters can be based on recent trends in birth and death rates. Similarly, changes in the carrying capacity of Earth's regions for the human economy can be estimated from recent or anticipated advances in technology and on predictions of environmental changes that may be caused by human activities. These types of information can be used to model future populations of human beings.

A typical prediction of recent population models is that the global abundance of human beings could reach about seven billion by 2011 or 2015, depending on birth rate variations, and 8.9 billion by 2050. Models tend to predict that the human population could stabilize at about 10–12 billion. in other words, the global abundance of human beings will probably double again before it stabilizes. This prediction is likely to be fulfilled unless there is an unpredicted, intervening catastrophe such as a collapse of the carrying capacity of the environment, an unprecedented and deadly pandemic, or a large-scale nuclear war.

Structure of Human Populations

Population structure refers to the relative abundance of males and females, and of individuals in various age classes. The latter type of structure is significantly different for growing versus stable populations and has important implications for future changes in population size.

Populations that have not been increasing or decreasing for some time have similar proportions in various age classes. In other words, there are comparable numbers of people aged five to 15 years old as those 35–45 years old. The distribution of people is even among age classes except for the very young and the very old, for whom there are, in many societies, disproportionately high risks of mortality. (In industrialized societies, the death rate for infants and young children may be very low; for the elderly, it remains high).

In contrast, populations that are growing rapidly have relatively more young people than older people. Therefore, the age-class structure of growing populations is triangular, that is, much wider at the bottom than at the top. for example, more than one half of the people in a rapidly growing human population might be less than 20 years old. This type of population structure implies inertia for further growth because of the increasing numbers of people that are continually reaching reproductive age.

Human populations that are growing rapidly for intrinsic reasons (i.e., birth rather than immigration) have a much higher birth rate than death rate and a markedly triangular age-class structure. The so-called demographic transition refers to the intermediate stage during which birth rates decrease to match death rates. Once this occurs, the age-class structure eventually becomes more equitable in distribution until zero population growth is achieved.

Environmental Effects of Human Population

The huge increases in size of the human population have resulted in a substantial degradation of environmental conditions. The changes have largely been characterized by deforestation, unsustainable harvesting of potentially renewable resources (such as wild animals and plants that are of economic importance), rapid mining of non-renewable resources (such as metals and fossil fuels), pollution, and other ecological damages.

At the same time that human populations have been increasing, there has also been a great intensification of per-capita environmental impacts. This has occurred through the direct and indirect consequences of increased resource use to sustain individual human beings and their social and technological infrastructure: meat production, fuel-burning, mining, air and water pollution, destruction of wild habitat, and so forth.

Demography

Demography is the study of human population with respect to size, composition, spatial distribution, and changes in the populations that occur over time.

Populations are never static, they grow or decline through the interplay of three demographic processes;

- Fertility

  Number of births that occur to an individual or a population.
- Mortality

  Death of an individual or the number of deaths that occur in a population.
- Migration

  The movement of people into or out of a specific geographic area.

The number of people in a given area can grow or decline as a result of the number of births that take place, thenumber of deaths that occur and/or by the number of people moving in or out of a locale. Migration can drastically change the size and composition of a population in a brief period of time, especially in small geographic locations.

Planners need to study changes in the composition of the population to plan for education, health care services, and economic development projects. The age of residents, gender, occupation, level of education, marital status, and living arrangements provide planners with the type of information needed to plan for the residents' diverse needs.

Knowing age ranges is of critical importance to planners since it is closely related to the demand for different types of services. A population composition that has a large percentage of residents under age 15 implies the need for schools, primary health services, and recreational needs. Gender is another key factor. For example, women, especially in their early reproductive years, 20–35, may need specialized health services for childbearing.

Ways Population Analysis is used in the Private and Public Sectors

Private Sector Planning

In the private sector (e.g., business) demographic information is equal to consumer information. Businesses use population characteristics, and demographic data to:

- Identify profitable locations for goods and services

  Selection of retail sites is usually based on population composition and the expected growth of its segments.
- Perform production analysis

  Determine characteristics of potential customers who may buy services or products. This analysis includes how the composition and location of the population are changing over time.
- Develop advertising strategies

Marketing products to different segments of the population such as youth, baby boomers, elderly, high income, etc.

- Perform strategic planning

  Tracking the growth or decline of existing markets, finding new markets and determining what new or existing products are most likely to be successful among different population groups.

The private sector is basically concerned about the "demand" and "location" for products and services. The composition or characteristics of the population is used as indicators of purchasing behavior. The business community also uses population information including estimates of population growth to identify the most profitable locations for goods and services in the future.

Public Sector Planning

Planners in the public sector use demographic information and analysis to assist with a number of planning decisions as indicated below.

- Planners use population information to determine the demand of services among different segments of the population. Demand is determined by the composition of the population and how it is changing over time — age-sex distribution, martial status, household types, occupation distribution, spatial distribution of the population, educational levels and income levels.
- Planners also study the present and future composition of the population and its spatial distribution to identify the best locations to provide services to meet local needs.
- Planners also examine population characteristics to determine the feasibility for new programs. A village may ask the government for a new school. Planners assess the age-sex distribution at present and in the future to determine whether or not its feasible to construct a new facility.
- Planners are also concerned about the impact of new plans on population change. A new plan to promote rural industries can lead to population growth as new families move into the community for job opportunities. Housing and educational plans may need to be revised to meet the needs of new households that may move into the area.
- Planners are also concerned about the impact of population growth on the ability to implement existing plans. Planners are interested in how changes in size, spatial distribution and composition will affect efforts to implement various types of plans — housing, social services, and infrastructure such as roads, water supply and electricity.

Use of Demographic Analysis in the Planning Process

Demographic analysis is needed in all stages of the planning process for both new and revised plans. Population analysis is needed to identify problems and community needs, establish goals and objectives, assess alternative courses of action, allocate resources for plan implementation, and evaluate the ability of the plan to achieve goals and objectives.

An example of how demographic analysis can be used in the planning process is presented in table. In the example, a health planner is asked to design a plan to expand primary health care within the district. The planner may use the following demographic analysis and information to develop the plan.

Table: Demographic Analysis in the Planning Process: Health Care Example

Planning Process	Demographic Analysis
Identify problems and needs	• Study trends in mortality rates and causes of death among different segments of the population • Study trends in fertility to plan for maternal and child health care • Project total population size by age-sex structure since it provides insights on the different health needs among different age groups
Goals and objectives	Collect information to establish objectives • Population size • Population composition • Geographic distribution • Population projections
Generating alternative strategies	Collect information on the size, location and composition of the target population to develop alternative strategies to achieve stated goals and objectives.
Select and implement a plan of action	Collect demographic data to determine: • Demand for services • Resource needs including the number of facilities, staff, medicine, and money • Locations for new facilities
Monitoring and evaluation	Use of demographic indicators to measure the achievement of the objectives. This could include • Infant mortality rates • Age specific fertility rates • Age specific death rates • Morbidity statistics

Human Population Control

Human population control is the practice of artificially altering the rate of growth of a human population. Historically, human population control has been implemented by limiting the population's birth rate, usually by government mandate, and has been undertaken as a response to factors including high or increasing levels of poverty, environmental concerns, religious reasons, and overpopulation. While population control can involve measures that improve people's lives by giving them greater control of their reproduction, some programs have exposed them to exploitation. Worldwide, the population control movement was active throughout the 1960s and 1970s, driving many reproductive health and family planning programs. In the 1980s, tension grew between population control advocates and feminist women's health activists who advanced women's reproductive rights as part of a human rights-based approach. Growing opposition to the narrow population control focus led to a significant change in population control policies in the early 1990s.

Methods

Population control may use one or more of the following practices although there are other methods as well:

- Contraception.
- Abstinence.
- Abortion.
- Emigration.
- Decreasing immigration.
- Starvation, famine.
- Pestilence, plague.
- War.

The methods chosen can be strongly influenced by the religious and cultural beliefs of community members. The failure of other methods of population control can lead to the use of abortion or infanticide as necessary final options. While a specific population control practice may be legal/mandated in one country, it may be illegal or restricted in another, indicative of the controversy surrounding this topic.

Ancient Times through Middle Ages

A number of ancient writers have reflected on the issue of population. At about 300 BC in India, Kautilya, a political philosopher, considered population as a source of political, economic, and military strength. Though a given region can house too many or too few people, he considered the latter possibility to be the greater evil. Kautilya favored the remarriage of widows (which at the time was forbidden in India), opposed taxes that encourage emigration, and believed that asceticism should be restricted to the aged.

In ancient Greece, Plato (427-347 BC) and Aristotle (384-322 BC) discussed the best population size for Greek city states such as Sparta, and concluded that cities should be small enough for efficient administration and direct citizen participation in public affairs, but at the same time needed to be large enough to defend themselves against hostile neighboring city states. In order to maintain a desired population size, the philosophers advised that procreation, and if necessary, immigration, should be encouraged if the population size was too small. Emigration to colonies would be encouraged should the population become too large. Aristotle concluded that a large increase in population would bring, "certain poverty on the citizenry, and poverty is the cause of sedition and evil." To halt rapid population increase, Aristotle advocated the use of abortion and the exposition of newborns.

Confucius (551-478 BC) and other Chinese writers cautioned that, "excessive growth may reduce output per worker, repress levels of living for the masses and engender strife." Confucius also observed that, "mortality increases when food supply is insufficient; that permanent marriage makes for high infantile mortality rates, that war checks population growth."

Ancient Rome, especially in the time of Augustus (63 BC- AD 14), needed manpower to acquire and administer the vast Roman Empire. A series of laws were instituted to encourage early marriage and frequent childbirth. Lex Julia (18 BC) and the Lex Papia Poppaea (AD 9) are two well known examples of such laws, which among others, provided tax breaks and preferential treatment when applying for public office for those that complied with the laws. Severe limitations were imposed on those who did not. For example, the surviving spouse of a childless couple could only inherit one-tenth of the deceased fortune, while the rest was taken by the state. These laws encountered resistance from the population which led to the disregard of their provisions and to their eventual abolition.

Tertullian, an early Christian author (ca AD 160-220), was one of the first to describe famine and war as factors that can prevent overpopulation. He wrote: "The strongest witness is the vast population of the earth to which we are a burden and she scarcely can provide for our needs; as our demands grow greater, our complaints against Nature's inadequacy are heard by all. The scourges of pestilence, famine, wars and earthquakes have come to be regarded as a blessing to overcrowded nations, since they serve to prune away the luxuriant growth of the human race."

Ibn Khaldoun, a famous North African Arab polymath (1332–1406), considered population changes to be connected to economic development, linking high birth rates and low death rates to times of economic upswing, and low birth rates and high death rates to economic downswing. Khaldoun concluded that high population density rather than high absolute population numbers were desirable to achieve more efficient division of labour and cheap administration.

During the Middle Ages in Christian Europe, population issues were rarely discussed in isolation. Attitudes were generally pro-natalist in line with the Biblical command, "Be ye fruitful and multiply."

The means of controlling and regulating populations can be traced back to many cultures. "Infanticide, or the killing of infants and young children, has occurred since early times. The Bible cities Abraham's intention to sacrifice his son, Isaac, to God. In early Rome, the father was given complete power to kill, abandon, or even sell his child. In Greek legend, Odepus was doomed to death until he was rescued by a family retainer. In Hawaii, China, and Japan, many female or disabled children were killed to maintain a strong race without overpopulation. Infanticide was practiced for many reasons. Like the Hawaiians, Chinese, and Japanese, some cultures saw the practice of infanticide as a means of controlling and regulating the population so that society's resources could be expanded on the strongest and most value (Crosson, Cynthia)."

16th and 17th Centuries

European cities grew more rapidly than before, and throughout the 16th century and early 17th century discussions on the advantages and disadvantages of population growth were frequent. Niccolò Machiavelli, an Italian Renaissance political philosopher, wrote, "When every province of the world so teems with inhabitants that they can neither subsist where they are nor remove themselves elsewhere the world will purge itself in one or another of these three ways," listing floods, plague and famine. Martin Luther, a German monk and theologian (1483–1546), concluded on the issue, "God makes children. He is also going to feed them." Jean Bodin, a French jurist and political philosopher (1530–1596), argued that a larger population would mean more production and in turn more export, which would increase the influx of silver and gold, and thus increase the riches of a country. Giovanni Botero, an Italian priest and diplomat (1540–1617), emphasized that, "the greatness of a city rests on the multitude of its inhabitants and their power," but pointed out that a population cannot increase beyond its food supply. If this limit was approached, late marriage, emigration, and war would serve to restore the balance.

Richard Hakluyt, an English writer (1527–1616), observed that, "Throughe our longe peace and seldome sickness wee are growen more populous than ever heretofore; many thousandes of idle persons are within this realme, which, havinge no way to be sett on worke, be either mutinous and seeke alteration in the state, or at leaste very burdensome to the commonwealthe." Hakluyt believed that this led to crime and full jails and in A Discourse on Western Planting (1584), Hakluyt advocated for the emigration of the surplus population. With the onset of the Thirty Year War (1618–1648), characterized by widespread devastation and deaths brought on by hunger and disease in Europe, concerns about depopulation returned.

The Population Control Movement

In the 20th century, population control proponents have drawn from the insights of Thomas Malthus, a British clergyman and economist who published *An Essay on the Principle of Population* in 1798. Malthus argued that, "Population, when unchecked, increases in a geometrical ratio. Subsistence increases only in an arithmetical ratio." He also outlined the idea of "positive checks" and "preventative checks." "Positive checks," such as diseases, war, disaster and famine, are factors that Malthus considered to increase the death rate. "Preventative checks" were factors that Malthus believed to affect the birth rate such as moral restraint, abstinence and birth control. He predicted that "positive checks" on exponential population growth would ultimately save humanity from itself and that human misery was an "absolute necessary consequence." Malthus went on to explain why he believed that this misery affected the poor in a disproportionate manner.

> "There is a constant effort towards an increase in population which tends to subject the lower classes of society to distress and to prevent any great permanent amelioration of their condition. The way in which these effects are produced seems to be this. We will suppose the means of subsistence in any country just equal to the easy support of its inhabitants. The constant effort towards population increases the number of people before the means of subsistence are increased. The food, therefore which before supplied seven millions must now be divided among seven millions and half or eight millions. The poor consequently must live much worse, and many of them be reduced to severe distress".

Finally, Malthus advocated for the education of the lower class about the use of "moral restraint," or voluntary abstinence, which he believed would slow the growth rate.

Paul R. Ehrlich, a US biologist and environmentalist, published The Population Bomb in 1968, advocating stringent population control policies. His central argument on population is as follows:

> "A cancer is an uncontrolled multiplication of cells; the population explosion is an uncontrolled multiplication of people. Treating only the symptoms of cancer may make the victim more comfortable at first, but eventually he dies - often horribly. A similar fate awaits a world with a population explosion if only the symptoms are treated. We must shift our efforts from treatment of the symptoms to the cutting out of the cancer. The operation will demand many apparent brutal and heartless decisions. The pain may be intense. But the disease is so far advanced that only with radical surgery does the patient have a chance to survive".

In his concluding chapter, Ehrlich offered a partial solution to the "population problem," "We need compulsory birth regulation though the addition of temporary sterilants to water supplies or staple food. Doses of the antidote would be carefully rationed by the government to produce the desired family size".

Ehrlich's views came to be accepted by many population control advocates in the United States and Europe in the 1960s and 1970s. Since Ehrlich introduced his idea of the "population bomb," overpopulation has been blamed for a variety of issues, including increasing poverty, high unemployment rates, environmental degradation, famine and genocide. In a 2004 interview, Ehrlich reviewed the predictions in his book, and found that while the specific dates within his predictions may have been wrong, his predictions about climate change and disease were valid. Ehrlich continued to advocate for population control and co-authored the book *The Population Explosion*, released in 1990 with his wife Anne Ehrlich. Paige Whaley Eager argues that the shift in perception that occurred in the 1960s must be understood in the context of the demographic changes that took place at the time. It was only in the first decade of the 19th century that the world's population reached one billion. The second billion was added in the 1930s, and the next billion in the 1960s. 90 percent of this net increase occurred in developing countries. Eager also argues that, at the time, the United States recognised that these demographic changes could significantly affect global geopolitics. Large increases occurred in China, Mexico and Nigeria, and demographers warned of a "population explosion," particularly in developing countries from the mid-1950s onwards.

Population Control and Economics

Opinions vary among economists about the effects of population change on a nation's economic health. Recent US scientific research concluded that the raising of a child costs about $16,000 yearly ($291,570 total for raising him/her up to his/her 18th birthday). In the USA, the multiplication of this number with the yearly population growth will yield the overall cost of the population growth. Costs for other developed countries are usually similar.

While some believe that reduction of the population is a key to economic growth, others argue that population reduction should be focused on what they judge to be undesirable sections of the population. Other economists doubt that a correlation between population reduction and economic growth exists. Some economists, such as Thomas Sowell and Walter E. Williams, have argued that poverty and famine are caused by bad government and bad economic policies, not by overpopulation. In his book, The Ultimate Resource, economist Julian Simon argued that higher population density leads to more specialization and technological innovation, which in turn leads to a higher standard of living. He claimed that human beings are the ultimate resource since we possess "productive and inventive minds that help find creative solutions to man's problems, thus leaving us better off over the long run". He also claimed that, "Our species is better off in just about every measurable material way." Simon also claimed that if you considered a list of countries ranked in order by population density, there is no correlation between population density and poverty and famine, and instead, if you considered a list of countries ranked in order by corruption within their respective governments, there is a significant correlation between government corruption and poverty and famine.

Contemporary Research

It is generally accepted that overpopulation is aggravated by poverty and gender inequality with consequent unavailability, and lack of knowledge of contraception. Third world evidence usually bears this theory out. However, first and second world fertility rates, in the Depression era United States, modern Russia, Japan, Italy, Sweden, Estonia and France suggest that these populations are responding inversely to poverty and economic pressures, especially on women. Thus, France is increasing social and women's services, like childcare and parental leave, expecting the policy to stop the aging of its population. Italy is regarded as alleviating overpopulation more rapidly than Sweden as a result of less gender equality and fewer children's services.

Newer research has been done by the U.S. National Security Council, in a study entitled National Security Study Memorandum 200, under the direction of Henry Kissinger in 1974. This report stressed that only 13 countries are projected to account for 47 percent of the world population increase by the year 2050. This, it is argued, (due to its impact on development, food requirements, resources and the environment) adversely affected the welfare and progress of countries concerned. It further argued that this would undermine the stability of countries friendly to the US and therefore harm the "national security" of the US as well.

David Pimentel, professor of ecology and agriculture at Cornell University, and Mario Giampietro, senior researcher at the National Research Institute on Food and Nutrition (INRAN), place in their study Food, Land, Population and the U.S. Economy the maximum U.S. population for a sustainable economy at 200 million. To achieve a sustainable economy and avert disaster, the United States must reduce its population by at least one-third, and world population will have to be reduced by two-thirds, says the study.

The authors of this study believe that agricultural crises will develop, but they will only begin to impact us after 2020, and will not become critical until 2050. Geologist Dale Allen Pfeiffer claims that coming decades could see spiraling food prices without relief and massive starvation on a global level never experienced before.

Another study has been done by the National Audubon Society which recently released a 16-page document called "Population and Habitat: Making the Connection." In this study, population control is widely supported.

Support for Population Control

Population control is also increasingly being featured in many environmental documentaries and films. An example is The Planet-documentary, which describes the ongoing rising human population, its effects on the planet, and the necessity of population control.

As early as 1798, Thomas Malthus stated in his Essay on the Principle of Population that population control needed to be implemented into society. Around the year 1900, Sir Francis Galton said in his publication called "Hereditary Improvement" that, "The unfit could become enemies to the State, if they continue to propagate." In 1968, Paul Ehrlich noted in The Population Bomb that, "We must cut the cancer of population growth," and that, "if this was not done, there would be only one other solution, namely the 'death rate solution' in which we raise the death rate through war-famine-pestilence etc." In the same year, another prominent modern advocate for mandatory population control was Garrett Hardin, who proposed in his landmark 1968 essay The Tragedy of the Commons that society must relinquish the "freedom to breed" through "mutual coercion, mutually agreed upon." Later on, in 1972, he reaffirmed his support in his new essay " Exploring New Ethics for Survival", by stating that, " We are breeding ourselves into oblivion." Other people advocating population control in the past were: Bertrand Russell, Margaret Sanger (1939), John D. Rockefeller, Frederick Osborn (1952), Isaac Asimov, Jacques Cousteau,

Today, a number of influential people advocate population control. They are:

- David Attenborough.
- Michael E. Arth.
- Jonathon Porritt, UK sustainable development commissioner.
- Sara Parkin.
- Crispin Tickell.

The head of the UN Millennium Project Jeffrey Sachs is also a heavy proponent of decreasing the effects of overpopulation. In 2007, Jeffrey Sachs gave a number of lectures about population control and overpopulation. In his lectures, called "Bursting at the Seams", he featured an integrated approach that would deal with a number of problems associated with overpopulation and poverty reduction. For example, when criticized for advocating mosquito nets he argued that child survival was, "by far one of the most powerful ways," to achieve fertility reduction, as this would assure poor families that the smaller number of children they had would survive.

Opposition to Population Control

The Roman Catholic Church has opposed abortion, sterilization, and contraception as a general practice, but specifically in regard to population control policies. Pope Benedict XVI has stated that "The extermination of millions of unborn children, in the name of the fight against poverty, actually constitutes the destruction of the poorest of all human beings".

Present-day Practice by Countries

China

The most significant modern population control system is China's one-child policy, in which, with various exceptions, having more than one child is discouraged. Unauthorized births are punished by fines, although there have also been allegations of illegal forced abortions and forced sterilization.

The Chinese government introduced the policy in 1978 to alleviate the social and environmental problems of China. According to government officials, the policy has helped prevent 400 million births. The success of the policy has been questioned, and reduction in fertility has also been attributed to the modernization of China. The policy is controversial both within and outside of China because of the issues it raises, the manner in which the policy has been implemented and because of concerns about negative economic and social consequences.

India

In India, only people with two or fewer children are eligible for election to a Gram panchayat, or local government. Indian people invented the Nirodh and decided to use contraception to control child birth.

We two, ours two ("Hum do, hamare do" in Hindi) is a slogan whose meaning is that of one family, two children and is intended to reinforce the message of population control.

Iran

Iran has succeeded in sharply reducing its birth rate in recent years. Iran is the only country where mandatory contraceptive courses are required for both males and females before a marriage license can be obtained. The government emphasizes the benefits of smaller families and the use of contraception.

United States

Enacted in 1970, Title X of the Public Health Service Act provides access to contraceptive services, supplies and information to those in need. Priority for services is given to persons of low-income. The Title X Family Planning program is administered through the Office of Population Affairs under the Office of Public Health and science. It is directed by the Office of Family Planning. In 2007, Congress appropriated roughly $283 million for family planning under Title X, at least 90 percent of which was used for services in family planning clinics. Title X is a vital source of funding for family planning clinics throughout the nation, which provide reproductive health care. The education and services supplied by the Title X-funded clinics support young individuals and low-income families. The goals of developing healthy families are accomplished by helping individuals and couples decide whether to have children and when the appropriate time to do so would be. Title X has made the prevention of unintended pregnancies possible. It has allowed millions of American women to receive necessary reproductive health care, plan their pregnancies and prevent abortions. Title X is dedicated exclusively to funding family planning and reproductive health care services.

Title X as a percentage of total public funding to family planning client services has steadily declined from 44% of total expenditures in 1980 to 12% in 2006. Medicaid has increased from 20% to 71% in the same time. In 2006, Medicaid contributed $1.3 billion to public family planning.

Population Policy

Population policy may be defined as deliberately constructed or modified institutional arrangements and/or specific programs through which governments influence, directly or indirectly, demographic change.

The generality of the definition lends itself to varying interpretations. For any given country, the aim of population policy may be narrowly construed as bringing about quantitative changes in the membership of the territorially circumscribed population under the government's jurisdiction. Additions to membership are effected only through births and immigration, losses are caused by emigration and by deaths. Concern with this last component is usually seen as a matter for health policy, leaving fertility and migration as the key objects of governmental interest in population policy.

More broadly, policy intent may also aim at modification of qualitative aspects of these phenomena–fertility and international migration–including the composition of the population by various demographic characteristics and the population's spatial distribution.

Furthermore, governments' concern with population matters can also extend beyond the borders of their own jurisdictions. International aspects of population policy have become increasingly salient in the contemporary world.

Population Control in Traditional Societies

Rulers of any political unit have a stake in the size and composition of the population over which they have authority, hence an incentive to try to influence demographic change in a desired direction. Thus "population policy" may be said to have a long history, starting at least with the empires of the ancient world. Greater numbers tended to connote greater wealth and power, at least for those at the apex of the social pyramid. Measures encouraging marriage and sometimes immigration testify to the prevailing populationist sentiment among rulers throughout history.

But the leverage of the weak premodern state over fertility in traditional societies was necessarily limited. The dominant influence setting the patterns of reproduction was located, instead, in a deeper layer of social interaction. Births, the key element affecting population change, are produced by individual couples–seemingly an intensely private affair yet one in which the immediate kin group and the surrounding local society in which that group is embedded have a material stake. All societies, if at varying degrees, grant a measure of self-sovereignty to their members. An individual has certain rights over his or her direction in life. But this is always subject to some constraints, not only biological but also social. Well before rights and obligations are formally codified in legal terms, they are established through spontaneous social interaction–a self-organizing process. Restrictions on freedom to act take the form of social expectations and pressures that individuals can ignore only at considerable personal costs to themselves. Typically, there is strong expectation that men and women should marry and have children. Parental and kin obligations in the matter of bringing up children are well understood by all adults and informally enforced by the community. In most societies there is the expectation that children are to be born to married couples only; that a man can have one wife at a time; that a husband is obligated to support his wife and a father his children; and that he can expect reciprocal services from them. And informal rules shaped by community interest tend effectively to regulate the entry of foreigners.

The fabric of such demographically relevant behavioral stances, supported by internalized personal norms and buttressed by religious injunctions, is a product of social evolution; how effective such institutions are becomes an important determinant of societal success. As a classic statement of the British demographer Alexander Carr-Saunders put it, persons and groups of persons:

> "Are naturally selected on account of the customs they practise, just as they are selected on account of their mental and physical characters. Those groups practicing the most advantageous customs will have an advantage in the constant struggle between adjacent groups over those that practise less advantageous customs. Few customs can be more advantageous than those which limit the number of a group to the desirable number. In the traditional society there would grow up an idea that it was the right thing to bring up a certain limited number of children, and the limitation of the family would be enforced by convention."

Given the harsh biological and economic constraints premodern societies invariably experienced, that "desirable number" presupposed fairly high fertility; high enough to provide a sufficient margin of safety over mortality. Successful societies–societies that survived to the dawn of the modern era–thus obeyed the biblical injunction to be fruitful and multiply, even though such multiplication as a matter of historical record was necessarily very slow. But traditional demographic regimes resulting from spontaneous social interaction achieved modest growth rates at varying levels of fertility and mortality. Early modern Western Europe succeeded in maintaining a relatively low average level of mortality by means of keeping birth rates low, primarily by means of a fairly high average age of marriage and substantial proportions that remained permanently single. A contrasting pattern, such as in India, combined early and universal marriage and a consequent high level of fertility with slow population growth by virtue of death rates that were also high, approximating the level of the birth rate. With respect to the rate of population growth these different combinations of birth and death rates in traditional societies were very similar. The potential for rapid population growth that might be triggered by a fall of mortality was, however much higher when the premodern equilibrium was the result of a combination of high mortality and high fertility.

Rationale for Population Policy

Modernity–the rise of democratic state formations reflecting the public interest and the emergence of rapid economic development–brought about the realistic promise of realizing age-old human aspirations for a better life. The state increasingly came to be seen as an institution created by the voluntary association of free individuals to further their interests. The central function of the state was to produce public goods–goods that individuals cannot secure for themselves. The U.S. Constitution, promulgated in 1789, articulated key items in the collective interest concisely and with universal validity. The aim of the Union formed by the People was, in the words of the Constitution's Preamble, to "establish Justice, insure domestic Tranquility, provide for the common defence, promote the general Welfare, and secure the Blessings of Liberty to ourselves and our Posterity." In pursuing such goals, regulation of immigration into a state's territory is clearly defined as a public good, thus delineating a particular role for population policy. And aggregate fertility may also be construed a public good, if its level as determined by spontaneous social interaction is too high or too low in terms of the collective interest.

The potential role of the state in regulating immigration is straightforward: individuals wishing to restrict or promote it cannot set up their own border patrols or issue entry visas. Individual preferences in the matter, however, are likely to differ. It is the task of the government to weigh and reconcile conflicting individual desires and come up with a policy deemed the best under the accepted rules of the political process.

To claim a role for the state in the matter of fertility is more problematic. Additions to the population are the result of a multitude of individual decisions concerning childbearing. Within the constraints of their social milieu, these decisions reflect an implicit calculus by parents about the private costs and benefits of children to them. But neither costs nor benefits of fertility are likely to be fully internal to the family: they can also impose burdens and advantages on others in the society. Such externalities, positive and negative, do represent a legitimate concern for all those affected. An individual's influence on the fertility of other families, however, is very limited: there are no private markets offering preferred patterns of aggregate demographic processes to individual buyers. Remedying such market failure may then be attempted through intervention by the state so as to affect individual behavior in order to best serve the common good–the good of all individuals.

The earliest clear formulation of the population problem as a problem of coordination among individual preferences, hence establishment of the rationale for potential state intervention in the matter of fertility, was given by William Foster Lloyd, an Oxford mathematician and economist, in an essay published in 1833. In the spirit of the Malthusian concerns of his time, Lloyd envisaged the possibility of overpopulation even under conditions when all families have only the children they actually want and suggested the direction in which remedy ought to be sought:

> "The simple fact of a country being overly populous is not, of itself, sufficient evidence that the fault lies in the people themselves, or a proof of the absence of a prudential disposition. The fault may rest, not with them as individuals, but with the constitution of society, of which they form part."

Population policy should therefore strive toward institutions and incentive systems–a constitution of society–that provide signals to individuals guiding them to behave in harmony with the collective interest.

Population Policy in the Liberal State

Technological progress and consequent improvements in the standard of living in modernizing societies result in a far more effective control of mortality than was possible in the traditional society. But the fall of the death rate accelerates the rate of population growth which, in turn, could strain the capacity of the economic system to accommodate the increased population numbers. Falling living standards then would once again increase death rates, reestablishing an approximate balance between births and deaths at a low standard of living. This was the pessimistic central vision of T. R. Malthus's 1798 Essay. But this outcome, although held to be highly probable, was, according to Malthus, avoidable. Given sound public policies, there was an alternative to subsistence-level equilibrium, both agreeable and achievable.

A salient element in the 1798 Essay, and in sub-sequent writings influenced by it, was disapproval of the schemes for poor relief prevailing in Britain and elsewhere in Europe–on the grounds that they were likely to encourage irresponsible reproduction. Efforts of the paternalistic state to reduce poverty were held to be misguided; by stimulating fertility, hence population growth, such efforts would generate only more misery. Malthusians argued that the state's correct stance in demographic matters, as in the economy at large, was laissez faire. This would foster the prudential habits among the general population similar to those that already existed among the propertied classes. It would do so by assuring that the costs of childbearing were not shared by society at large but were primarily borne by the individual couples having children.

Heeding such a prescription did not imply that the state was to play a passive role in demographic matters. Malthus's own writings, most clearly his 1820 tract "Principles of Political Economy", spell out a broad agenda which expresses the philosophy that came to be dominant in the liberal states of the West in the nineteenth century. Material improvements, such as higher wages for labor, could indeed be defeated if they would be "chiefly spent in the maintenance of large and frequent families." But Malthus also envisaged a different, happier possible outcome: "a decided improvement in the modes of subsistence, and the conveniences and comforts enjoyed, without a proportionate acceleration of the rate of (population) increase."

The possibility of such diametrically different responses to the stimulus of higher wages suggests a large element of indeterminacy in fertility behavior. To Malthus, the causes of these divergent responses were to be found in the circumstances, social and political, in which people lived–in particular, whether those circumstances hindered or rewarded planning for the future. From his analysis he derived a prescription for a population policy that would yield the hoped-for demographic outcome:

> "Of all the causes which tend to generate prudential habits among the lower classes of society, the most essential is unquestionably civil liberty. No people can be much accustomed to form plans for the future, who do not feel assured that their industrious exertions, while fair and honourable, will be allowed to have free scope; and that the property which they either possess, or may acquire, will be secured to them by a known code of just laws impartially administered. But it has been found by experience, that civil liberty cannot be secured without political liberty. Consequently, political liberty becomes almost equally essential."

During the long nineteenth-century–that may be thought of as stretching to the outbreak of the First World War–the politics in Europe and in its over-seas offshoots favored, even if imperfectly, the development of institutional and legal frameworks in harmony with such principles. This, in interaction with economic and cultural changes shaped by the industrial revolution, created a milieu that fostered the prudential habits of parents, rendering the micro-level calculus of the costs and benefits of children increasingly salient. Rising demand for labor, including greater use of child labor, and rising income levels tended to sustain high fertility or even to stimulate it. But rising material expectations, broadening opportunities for social mobility, and the patterns and circumstance of urban living pulled in the opposite direction. This was powerfully reinforced by some programmatic activities that were consistent with the limited role the liberal state claimed in managing the economy. These included public health programs and projects aimed at improving basic infrastructure for transport and communication.

And most importantly, the state, or local government, assumed a key role in fostering, organizing, and financing public education. At basic levels school attendance was made mandatory and enforced and, in parallel, labor laws curtailed the employment of children.

Reflecting long-standing cultural values and religious injunctions, and contrary to laissez-faire principles, the liberal state generally banned the spreading of contraceptive information and the sale of contraceptive devices and made abortion illegal. Such restrictions typically remained in effect well into the twentieth century. But by all evidence, any upward pressure on fertility from these restrictions was swamped by the downward pressure on parental demand for children resulting from the state policies and programs just mentioned. By the last quarter of the nineteenth century birth rates were falling rapidly in the countries of the West. In many cases, rates of population growth fell also, despite continuing improvements in mortality. In Europe this trend was facilitated by emigration, which both sending and receiving countries–notably the United States, Canada, and Australia–either positively encouraged or at least permitted.

The stance of the liberal state on population policy thus brought about the prospect of a new demographic equilibrium in the West that could be consistent with continuing material progress: achievement of a stationary population at low levels of fertility and mortality and allowing freedom of movement internationally.

Population Policy between the World Wars

The massive losses of life resulting from World War I and from the influenza pandemic at its immediate aftermath, and the sharp drop in the number of births during the war years, were temporary disruptions in the steadily declining trends of fertility and mortality characterizing the prewar decades in the West. Those trends soon made it evident that there is no built-in guarantee that the sum total of individual fertility decisions will eventually settle at a point at which, in the aggregate, the rate of population growth will be exactly zero or fluctuate tightly around a zero rate. Although, owing to relatively youthful age distributions, the rate of natural increase remained positive, by the late 1920s demographers realized that fertility rates in several Western countries had fallen to such a low level that, in the longer term, natural increase would become negative. This trend became more accentuated and more general under the impact of the Great Depression. Some observers foresaw a "twilight of parenthood."

Just as excessive reproduction called for corrective public policies, there were calls for corrective action achieving the opposite result: enhancing fertility so as to assure at least the simple maintenance of the population. In some countries the ban on contraceptives was tightened and the penalties on abortion were increased. These measures had little effect. So did, predictably, governmental exhortation appealing to families to have more children.

The most promising avenue for population policy seemed to be to use the instruments available to the state for redistributing income so as to reward demographic behavior considered socially desirable (and to discourage contrary behavior). By the 1930s such pronatalist policies came to be fairly widely if rather tight-fistedly applied in a number of countries. Among Europe's emerging democratic welfare states Sweden and France were pioneers in providing financial rewards and services in kind to families with children, especially to larger families. (Sweden, however, also allowed liberal access to contraception.) Similar policies were applied with equal or greater vigor in fascist Italy and Germany.

Invariably, the proponents of such policies claimed some results in terms of birth rates somewhat higher than would have been expected in their absence. But the latter quantity is a hypothetical one, which introduces a necessary caution to such claims. More pertinently, when average fertility is low, the birth rate in any given year is an unreliable measure of long-run fertility. Couples have considerable latitude to time the birth of their children earlier or later, without affecting the number of children they ultimately wish to have. Logically, pronatalist policies seek to affect that lifetime total rather than aiming at temporary increases in the birth rate.

International Population Policy after World War II

In the countries that the United Nations categorizes as less developed, population policy issues attracted little attention until the middle of the twentieth century. Fertility remained high, more or less at its pre-modern level. Population growth was slowly increasing, however, as a result of improvements in mortality. Following the end of World War II, mortality decline accelerated greatly and as a result so did population growth. In 1950 the world population was 2.5 billion. Some 1.7 billion of that total was in countries classified as less developed, with an average annual birth rate of 44 per 1000 population–twice as high as in the more developed group. Unless a decline of the birth rate got under way fairly rapidly, an unprecedentedly large expansion of human numbers was inevitable.

Although the trigger of such population growth was a welcome development–falling death rates–growth rates that would double or even treble a population within a generation seemed a major obstacle to development. And the large and widening differential between the more developed and the less developed countries in terms of population size and average income levels was seen as holding out the prospect of major dislocations and long term instabilities within the international system. As a result, in the 1950s an intense debate started on what policies could reduce fertility in the less developed world. This policy debate was primarily Western, much of it American, just as the diagnosis of the problem itself had been. The proposals that emerged were to be applied in countries representing a large and growing share of the global population. Population policy became international.

The West of course already had relatively low fertility, and with it much historical experience on why birth rates fell. Social science analysis was virtually unanimous in interpreting this experience. The explanation centered on the role of changing structural conditions of the economy, conditions to which micro-level units of the population tended to respond, in demographic as well as in other matters. Demand for smaller families was seen as the primary force determining birth rates; the means by which couples regulated their fertility was not unimportant, but seen as a distinctly secondary factor. If the demand was strong enough, fertility would be low, even if birth control technology was primitive. A transition to low fertility presupposed changing preferences, and such preferences were responses to market signals. If policy was to have an explicit role, it would be through reinforcing those signals.

In the experience of past fertility transitions four components of the incentive structure seemed especially pertinent:

1. The direct costs parents must incur in bringing up children;
2. The opportunity costs of children to parents; that is, the earnings a couple must forgo because of children;
3. The contribution of children to family income through labor services;
4. The contributions of children to parents' economic security in old age, in comparison to alternative sources of security.

Fertility declines when shifts in these components make family limitation advantageous to couples, overcoming cultural resistance supporting traditional behavior. Patterns of development generate that effect when at least some, but especially when all of the following conditions are fulfilled:

1. Social expectations and formal institutional arrangements place on parents the major financial responsibility for raising their own children, including much of the cost of education and health care;
2. Women have access to income-earning opportunities in the labor market, including jobs not easily compatible with childbearing and childrearing;
3. Social institutions make formal education (primary and early secondary) compulsory and effectively enforce school attendance;
4. Child labor is made illegal;
5. Effective legal guarantees of property rights, legal enforcement of private contracts, and the development of public and private insurance and pension schemes provide attractive and comparatively secure alternatives to children as a source of old-age security.

Social and institutional conditions that make such changes potent generators of fertility decline include the following:

1. Emphasis on personal economic contribution (rather than, for example, class status or political loyalty) as the primary factor determining a person's earnings, thus providing an incentive for increased investment in human capital;
2. Systems of promotion that provide opportunities for upward social mobility according to merit and tolerate downward social mobility;
3. Openness to outside influences that create rising expectations with respect to material standards of living;
4. Emphasis not only on the rights but also on the social and economic responsibilities of the individual.

Some of the factors that prompted the fall of fertility in the West also became potent in the less developed countries as concomitants of successful economic and social progress. International conferences, for example the 1974 Bucharest conference on population, spelled out many of the essential socioeconomic changes necessary for fertility to shift from high to low levels. But assembling the instruments so identified into a coherent development strategy of institutional-structural reforms, reforms justified also by demographic objectives, remained an elusive task. In promoting development governments came to see their roles not in supporting institutions harnessing the market, but replacing the market in key developmental tasks through specific goal-oriented categorical programs. In the matter of population policy, the rapid postwar progress in the technology of birth control provided an appealing apparent short-cut for achieving fertility decline through programmatic means. Markets, it was held, could not be relied on to bring that technology to those wishing to practice birth control. Governments could, instead, organize free delivery of birth control information, and provide effective means for preventing births to all those, primarily (it was assumed) women, who wished to plan their families. Surveys indicated that there existed a substantial latent demand for such services. Satisfied customers, in turn, would serve as role models, bringing new clients to the program.

By the mid-1960s, in programmatic terms the international population policy debate on the relative importance of demand versus supply was essentially decided in favor of the latter. For the next quarter century, population policy in the developing world became essentially synonymous with family planning programs.

Financial and administrative limitations within developing countries necessitated the heavy involvement of foreign assistance in launching and sustaining family planning programs. Although donor countries' own domestic experience in this area was practically nonexistent, such aid, justified by the seriousness of the "population problem" (a problem whose cause was defined in terms of aggregate indexes of population growth) was readily forthcoming, partly in the form of bilateral assistance and partly through international aid institutions.

The cost of birth control technology was, in itself, relatively modest. But sustaining an effective delivery service did represent significant claims on scarce human and material resources. Declared demand for birth control does not necessarily translate into effective willingness to practice it: conflicting desires may interfere. Weakness of measures of latent demand, or "unmet need," is reflected in the requirements that programs are supposed to satisfy if they are to be successful. These typically included such items as "doorstep accessibility of quality services," "broad choice of contraceptive methods," "forceful IEC (information, education, and communication) programs," "sound financing strategies," "sound management with proper logistics," "evaluation systems," "a continuous process of strategic thinking, planning and management," and "staff leadership for program parameters".

The effectiveness of family planning programs in reducing fertility remains a matter of controversy. According to international guidelines, programs recruit their clients on a strictly voluntary basis. By accepting the service voluntarily, the individual acceptor demonstrates that she values that service. But some of the more successful programs, notably in Asia, tended to increase acceptance by often heavy-handed methods of persuasion, and, in the especially important case of China, by coercion backed by legal sanctions. Where fertility fell in less developed countries with active family planning programs, it is typically found that program-provided free services account for a large percentage of those practicing contraception. This non-surprising result is then often taken as an indicator of success in reducing aggregate fertility. But what would have happened in the absence of the program is conjectural, hence routinely ignored. Some less developed countries that lacked government programs also experienced major falls of fertility: Brazil is a conspicuous example. Similarly, if programs have seemingly only minor success in reducing fertility, this can be taken as evidence that the program is inadequately financed, organized, and managed: greater efforts would have led to better results.

Family planning programs as they were commonly conceived had a strong resemblance to health programs. But given the special priority accorded to family planning services in foreign assistance, typically they were organized as a separate "vertical" program, or kept administratively distinct within the broad health program. The justification for such treatment was that while acceptors of family planning services are recruited because the program satisfies their individual need, the program also serves a national developmental need by helping to reduce aggregate population growth, hence deserves priority. Once a family planning program is organized, its managerial and professional cadres form a natural advocacy group strongly interested in the program's sustenance. Invoking the public interest in lowered fertility, as distinct from simply serving the needs of the clients of the program, has long served as a key supporting argument in that endeavor.

Over time, this developmental prop has eroded. This was in part a result of criticisms of the intrinsic scientific merit of the argument but most of all a reflection of the extensive decline of fertility that has occurred, a decline often attributed to the success of the family programs themselves. At the eve of the 1994 International Conference on Population and Development, a review, considered highly respectful of the importance of fertility decline for successful development, reached the guarded conclusion that the evidence on the subject "mostly support the view that rapid population growth in poor countries under conditions of high fertility is inimical to many development goals"–with stress on the qualifying words "mostly," "high fertility," "many development goals," and "rapid," and with the last-mentioned term defined as "in excess of 2 percent annually". By that time, among world regions, only Africa and West Asia had a population growth rate meeting that criterion.

Accordingly, the development rationale of family planning programs was gradually dropped and was replaced by the argument that the programs satisfy important health needs and help people exercise a fundamental human right. The Cairo conference formalized this shift: even though the name of the conference for the first time included a reference to development, scant attention was paid to that concept. Family planning programs were redefined, instead, as reproductive health programs, responding to a broader range of women's health needs, such as prevention of unsafe abortions and sexually transmitted diseases, including HIV/AIDS. But beyond this, new emphasis was put on some requirements that would contribute to women's empowerment: reduction of infant and maternal mortality and improvement in girls' education and women's opportunities for employment and political participation. Although the connection was not highlighted, these are conditions that are likely to help reduce the birth rate through stimulating the demand for smaller family size. The Cairo conference, in effect, reverted to some key elements of a demand strategy for reducing birth rates.

The future of family planning service programs are thus left in a somewhat tenuous status. Without invoking a collective interest in a wider practice of birth control, it is not clear what level of priority should be accorded to such programs as just one part of publicly financed health programs, or indeed relative to many other social welfare programs that also serve demonstrable human needs. Not surprisingly, there are increasing efforts in national programs to rely on the market in enhancing access to contraceptives and to provide program services on a fee-for-service basis.

Population Policy in Response to Below Replacement Fertility

During the second half of the twentieth century debates about population policy, and consequent programmatic action, were centered on the issue of rapid population growth in the less developed world. Toward the end of this period, however, a quite different demographic phenomenon has begun to attract increasing attention: aggregate fertility levels that are inadequate for the long-run maintenance of the population. Analytically, the potential population policy issue raised by low fertility is identical to the problem inherent in rapid population growth: it is caused by the disjunction between the sum total of individual reproductive decisions and the collective interest in a long-run demographic equilibrium. But this time individual aspirations generate a deficit rather than an excess in population growth. The syndrome, is not entirely novel: it was detectable in fertility trends in the West, especially in Europe, in the 1920s and 1930s, and in some instances, notably in France, even earlier. But in the decades immediately following World War II the baby boom seemed to make the issue of low fertility moot. Indeed, by any historical standard, population growth was rapid during the second half of the twentieth century even in the developed world. Europe's population, for example, grew during that period from 550 million to about 730 million.

The baby boom was, however, a temporary interruption of the secular downward trend in fertility. By the 1970s the net reproduction rate was at or below unity in most countries in Europe and also in the United States. In the U.S. fertility stabilized at or very close to that rate, but in Europe fertility continued to decline. By the beginning of the twenty-first century, the average total fertility rate was 1.4. Such a level, if maintained indefinitely, would result in a population loss of one-third from generation to generation, that is, roughly, over each period of some 30 years. Some countries, notably in Southern, Central, and Eastern Europe, period fertility rates were at low levels without historical precedent for large populations. If continued, in the absence of large compensatory immigration this would not only lead to rapid population decline but also result in very high proportions of the population at old ages. It might be expected that in the affected countries such prospects would generate not only concern, but also vigorous remedial policy action.

By and large, however, this response has not been evident. Most governments as well as the general public tend to view below-replacement fertility with an equanimity quite unlike the alarmed reaction that the same phenomenon elicited when it first emerged between the two World Wars. And explicit pronatalist policies, common in the 1930s, are conspicuous by their absence. There are a variety of reasons explaining this indifference.

First, the preeminent population issue confronting policymakers in the post-World War II period was rapid global population growth. Programs aimed at moderating fertility in the developing world received assistance or at least encouragement from the rich, low-fertility countries. Although the rationale was modified over time, such assistance and encouragement has continued, as indeed substantial further population increase in the less developed countries is still anticipated in the early decades of the twenty-first century. Even though population issues tend to be sui generis, reflecting differences in demographic behavior country-by-country, there was, and remains, a perceived dissonance between fertility-lowering assistance to other countries and engaging in action at home serving the opposite aims. Faulty logic notwithstanding, the international terrain has not been favorable for domestic pronatalism.

Second, the natural rate of increase–the difference between the number of births and the number of deaths–is still positive in many of the countries with fertility well below replacement. This is the result of age distributions that reflect past fertility and mortality, and notably the effects of the postwar baby boom, that still favor population growth. While this momentum effect is temporary, the longer-term implications for population decline and population aging are only dimly perceived by the general public and provide an excuse for inaction on the part of policymakers.

Third, when those longer-term demographic effects are understood, a calmer attitude still prevails. There is an inclination, reinforced by increasing concern with the quality of the natural environment, to regard a degree of demographic "decompression" as a not necessarily unwelcome prospect, especially in countries with an already dense population. And it is assumed that the economic and social disadvantages that might be imposed by a declining population can be effectively dealt with through institutional adjustments and social policy measures other than measures aiming for a higher birth rate. A demographic policy often regarded as potentially helpful in this regard is encouragement of immigration. That willing immigrants are available to compensate for low birth rates is taken for granted–a realistic assumption in high-income countries.

Fourth, there is a vague expectation that the population decline, impending or already begun, will in due course trigger corrective homeostatic mechanisms, leading to a spontaneous rebound in the level of fertility. Another baby boom may not be in the offing, but fertility may rise sufficiently to once again reach or at least approximate replacement level. Governments, it is assumed, would be illadvised to interfere with this natural process by trying to increase birth rates and then seek to fine-tune them at the desirable steady-state level. According to this view, a laisser-faire fertility policy is justified since, apart from broad upper and lower limits, governments are not competent to determine what constitutes an optimal fertility rate, or growth rate, or population size in any given year, decade, or even longer time interval.

Finally, even if the will were there, there is a paucity of effective pronatalist policy instruments. Exhortation from governments are not promising, and in any case unlikely to be tried in a democratic polity. Restrictions imposed on access to modern contraceptive technology are not politically acceptable; they would be also certain to prove a failure. This leaves the traditional levers of social policy: dispensing material incentives and disincentives so as to increase the willingness of couples to have children. Such incentives can be engineered by the government through fiscal measures, such as differential taxation, and/or through provision of services in kind. This approach was tried in the interwar years, but, with at best limited success. After World War II many similar measures continued to be applied; in fact, with the steadily expanding welfare state, they were often upgraded and their scope, too, was extended. However, they were no longer considered "pronatalist" but were absorbed within the more encompassing frames of family and general welfare policy. The new label partly reflected a political-ideological preference, but in part also the fact that some distinctive features of pronatalism–such as differential rewards that favored large families, and non-means-tested or even regressive allocation of family and child benefits–were generally no longer acceptable.

Although the redistributive policies of the contemporary welfare state are biased in favor of the elderly and the poor, government-organized transfers to parents of children, or to children directly (such as through publicly financed day-care services and free or subsidized education often beyond the secondary school level, which lessen the cost of children to parents), are substantial in all low fertility countries. Indeed, it is typically assumed that existing family and welfare policies sustain fertility above a level that would ensue in their absence. Accordingly, making these policies more generous–socializing an even larger share of child costs–is often seen as a means toward increasing fertility, whether as an outright policy objective or, more in the prevailing spirit of the time, as an unintended but welcome byproduct. Such extension, however, is difficult, given the fiscal constraints of already overcommitted welfare states. And more to the point, the net effect of family-friendly redistribution of incomes and provision of services is uncertain. It is notable that in the United States, where such schemes are distinctly less well funded than, for example, in Western Europe, fertility is, nevertheless, relatively high.

In recent decades, in modern industrial economies, participation of women in the formal labor force expanded rapidly. This tendency, reflecting market forces but also encouraged by government policy (partly as an antidote for deteriorating dependency ratios as the population becomes older), is likely to continue. Among the factors explaining the low level of fertility despite general material affluence, many observers point to the double burden on women of both raising children and working outside the home. To the extent that higher birth rates are seen to be socially desirable, the derived policy prescription is to adopt measures that make motherhood and women's labor force participation more compatible. The higher fertility in countries (notably in Scandinavia) where such measures are strongly applied, compared to countries (especially those in Southern Europe) where they are largely absent, suggests that enhanced compatibility (through day-care services, flexible work-hours, liberal sick-leave allowances, and the like) is an effective pronatalist policy even if motivated by other considerations. But it is far from clear whether the fertility differential so generated is high enough to bring the total fertility rate back to replacement level. Steady labor force participation of women during the childbearing years can certainly be made compatible with having one child or even two. It is likely to be far less compatible with sustaining, or even increasing, the proportion of women who have more than two children. Many career-oriented women voluntarily remain childless; many others prefer a single child. It follows that, to achieve average replacement-level fertility, the proportions of such women need to be counterbalanced by high enough proportions of women who have chosen third-, fourth-, or even higher-order births. There is little indication at present that policies directed at enhanced compatibility achieve that result.

When fertility is high, as it still is in most developing countries, it is a safe prediction that with economic development it will eventually decline, at least to replacement level. In the longer term, apart from outmigration, the only alternative is higher mortality.

The record of the high-income countries indicates, however, that replacement level fertility is not a necessary resting point. Once fertility is lower than that, predictions become highly hazardous. European, and also East Asian experience suggests that fertility has a tendency to settle below an average of two children per woman, hence a tendency toward sustained population decline. The question, to which no good answers exist at the dawn of the twenty-first century, is "how far below?" "Scandinavian"-style family policies may stabilize fertility only modestly below replacement–such as around a total fertility rate of 1.8. That would imply a fairly moderate relative shortfall of births compared to deaths, and population stability in rich countries with such vital rates could be fully or nearly compensated with a modest level of controlled immigration. Population aging would be then kept within relatively narrow limits, which postindustrial economies could readily adjust to. The demographic weight of such countries within the global total in the foreseeable future would continue to shrink, raising possible problems of a shifting geopolitical balance. Still, such demographic configurations would be likely to push the day of demographic reckoning beyond the policy horizons that governments feel an obligation to be actively concerned with. Pronatalist interventions would find at best a marginal place on governments' policy agendas.

On the other hand, fertility levels in the lowest fertility countries–countries with a total fertility rate of 1.3 or below around the turn of the century–might stabilize at that level, or even shrink further, reflecting the decentralized and uncoordinated decisions of individuals and individual couples. Such an outcome might also foreshadow future reproductive behavior in countries in which fertility is still fairly close to replacement level. This would create a qualitatively different demographic situation for which there are few precedents in modern history. It would represent a clear threat to the continuing viability of the countries affected. Compensatory immigration flows would have to be so large as to be inconsistent with any reasonable degree of cultural and ethnic continuity. Alternatively, population aging in the absence of immigration would create virtually unsolvable challenges, and there would be a likely drastic loss of relative geopolitical status. Spontaneous homeostatic mechanisms may not come into play to save the day, or may do so too sluggishly to matter. A radical rethinking of fertility policy would then become a necessity for social–and national–survival.

Mortality Rate

Mortality, in demographic usage, the frequency of death in a population.

In general, the risk of death at any given age is less for females than for males, except during the childbearing years (in economically developed societies females have a lower mortality even during those years). The risk of death for both sexes is high immediately after birth, diminishing during childhood and reaching a minimum at 10 to 12 years of age. The risk then rises again, until at late ages it surpasses that of the first year of life. The expectation of life at birth is the most efficient index of the general level of mortality of a population. In ancient Greece and Rome the average life expectancy was about 28 years; in the early 21st century life expectancy averaged about 78 years in most industrialized countries. In countries with a high rate of HIV infection, however, the average life expectancy was as low as 33 years.

A mortality rate is a measure of the frequency of occurrence of death in a defined population during a specified interval.

The formula for calculating the mortality rate is as follows:

$$\text{Mortality Rate} = \frac{\text{number of deaths in a given time period}}{\text{population from which the deaths occurred in the same time period}} \times 10^{n}$$

When the mortality rate is based on vital statistics (e.g. counts of birth or death certificates), the denominator is usually the population at the mid-point of the given time period.

Infant Mortality Rate

Infant mortality rate, measure of human infant deaths in a group younger than one year of age. It is an important indicator of the overall physical health of a community. Preserving the lives of newborns has been a long-standing issue in public health, social policy, and humanitarian endeavours. High infant mortality rates are generally indicative of unmet human health needs in sanitation, medical care, nutrition, and education.

The infant mortality rate is an age-specific ratio used by epidemiologists, demographers, physicians, and social scientists to better understand the extent and causes of infant deaths. To compute a given year's infant mortality rate in a certain area, one would need to know how many babies were born alive in the area during the period and how many babies who were born alive died before their first birthday during that time. The number of infant deaths is then divided by the number of infant births, and the results are multiplied by 1,000 so that the rate reflects the number of infant deaths per 1,000 births in a standardized manner. Alternately, the rate could be multiplied by 10,000 or 1,000,000, depending on the desired comparison level.

There are a number of causes of infant mortality, including poor sanitation, poor water quality, malnourishment of the mother and infant, inadequate prenatal and medical care, and use of infant formula as a breast milk substitute. Women's status and disparities of wealth are also reflected in infant mortality rates. In areas where women have few rights and where there is a large income difference between the poor and the wealthy, infant mortality rates tend to be high. Contributing to the problem are poor education and limited access to birth control, both of which lead to high numbers of births per mother and short intervals between births. High-frequency births allow less recovery time for mothers and entail potential food shortages in poor families. When women are educated, they are more likely to give birth at later ages and to seek better health care and better education for their children, including their daughters.

Poor Sanitation and Water Quality

In least-developed countries (LDC) a primary cause of infant mortality is poor quality of water. Drinking water that has been contaminated by fecal material or other infectious organisms can cause life-threatening diarrhea and vomiting in infants. A lack of clean drinking water leads to dehydration and fluid volume depletion. The loss of large quantities of fluids and salts from the body can quickly kill an infant. Adequate clean water must also be available for hygiene to maintain the health of infants. Advocacy groups estimate that the deaths of several million children yearly could be prevented by the use of a simple low-cost oral rehydration solution.

Breastfeeding Controversies

The use of infant formula has come under attack in both developing countries and LDCs as well as in the industrialized world. Many forms of infant formula start as powders that must be mixed with water to be used. The World Health Organization (WHO) has questioned the use of breast milk substitutes in poor families, particularly in areas where clean water is not available, because it may increase the risk of infant death.

In the 1970s the Nestlé Corporation was criticized by a number of groups for its distribution of infant formula to women in developing countries. The company distributed free samples of infant formula and marketed them to women as a more modern Western alternative to breast milk. Unfortunately, many women did not realize that their breast milk production would decrease or stop entirely when they started to rely on infant formula. Without breast milk, they had little choice but to continue to use the formula, often with disastrous results. Because the formula was expensive, poor families tried to stretch their supply by watering it down. That practice led to malnutrition, starvation, or digestive infections in the frequent cases in which the diluting water was not clean. In addition, because breastfeeding stimulates hormones that serve as a semieffective natural contraceptive, that benefit was lost with the switch to artificial infant formulas.

Low Birth Weight

Low birth weight is the single most significant characteristic associated with higher infant mortality. In industrialized countries, low birth weights are characteristic of premature births. However, in LDCs they more frequently occur at full term, because of a lack of adequate maternal nutrition or because of malaria, measles, or other infectious diseases, such as HIV. For a full-term infant, low birth weight is a weight less than 2,500 grams (5 pounds 8 ounces) at birth or a weight that is one standard deviation or more below the weight expected for that age in a reference population.

Chances of survival for a premature birth vary greatly depending on the available resources. Premature infants (born at less than 37 weeks gestation) have a higher risk of death not only because of low birth weight but also because their respiratory and digestive systems are not fully mature.

Prenatal Care

Good prenatal care has been linked to reduced infant mortality. Ideally, prenatal care should begin as early in the pregnancy as possible, with visits to a health care provider every 4 weeks during the first 28 weeks of pregnancy, every 2 to 3 weeks for the next 8 weeks, and weekly thereafter until delivery. Using this formula, the appropriate number of prenatal visits for a 39-week pregnancy is 12. Maintaining such a schedule requires time, effort, and—most important—access to a system of affordable health care, which is lacking in many LDCs and even in some industrialized countries.

Race and Infant Mortality

The United States may be regarded as having several infant mortality rates: one for the white population, one for the Asian population, one for Native Americans, and another for African Americans. African American babies are nearly two and a half times more likely to die before their first birthday than are white babies. According to the U.S. Center for Disease Control, infant mortality among African Americans in 2014 occurred at a rate of 10.93 deaths per 1,000 live births, nearly twice the national average of 5.82. Preterm delivery with low birth weight is the leading cause of infant death for African American infants, occurring at a rate that is more than three times greater than that for white Americans. Sudden infant death syndrome (SIDS) is more than twice as common among American Indian and Alaska Natives than it is among non-Hispanic whites. Those differences stem from differences in health access, poverty, and other effects of racism. Rates of infant mortality among Chinese and Japanese Americans tend to be the lowest of all U.S. ethnic groups.

Crude Death Rate

Crude death rate is the total number of deaths to residents in a specified geographic area (country, state, county, etc.) divided by the total population for the same geographic area (for a specified time period, usually a calendar year) and multiplied by 100,000.

A death rate is defined as,

$$\text{death rate} = \frac{\text{number of deaths in a specified time period}}{\text{number of people exposed to the risk of dying during that time period}}$$

Thus, in order to measure mortality, data are required about the number of deaths, and about the number of people exposed to the risk of dying. Data on the number of deaths are usually obtained from death registers, and data on the number of people exposed to the risk of dying are typically obtained from a population census. Of course, survey data may also be used, especially in countries where death registration is deficient, or the quality of census data is suspect.

The simplest conceivable death rate is probably the total number of deaths in a given time period divided by the total population. This measure is called the crude death rate. The time period used is typically one calendar year. Thus,

$$\text{crude death rate} = \frac{\text{total number of deaths in a given year}}{\text{total population}}.$$

An immediate issue arises with the measurement of the total population. During any year, the population will usually change. At what point in the year, therefore, should it be measured? Conventionally, the point chosen is half-way through the year (30 June). The population on 30 June is called the mid-year population. Using this definition of the population exposed to the risk of dying, therefore,

$$\text{crude death rate} = \frac{\text{total number of deaths in a given year}}{\text{total mid- year population}}.$$

Denoting the crude death rate in year *t* by the symbol d_t, the total number of deaths in year *t* by θ_t and the total population on 30 June in year *t* by p_t we can write,

$$d_t = \frac{\theta_t}{p_t}$$

Now, for simplicity, the subscripts t are usually omitted because, unless otherwise stated, the period of time over which the crude death rate is measured may be assumed to be a single calendar year. Thus,

$$d = \frac{\theta}{p}.$$

Since death is a relatively rare event in most populations, the crude death rate is often small. For this reason, it is often expressed as the number of deaths per thousand of the population, or

$$d = \frac{\theta}{p} \times 1000.$$

Thus, for example, the population of Peru on 30 June 1989 has been estimated to be 21

113000 (excluding some Indian people in remote areas). It is estimated that there were 200468 deaths in Peru in 1989. The crude death rate in Peru in 1989 is therefore equal to 200468/21 113000, which is 0.00950, or, multiplying by 1000, 9.5 per thousand.

Age-specific Death Rates

The crude death rate does not provide a great deal of information about mortality. In particular, the risk of dying varies greatly with age, and the crude death rate indicates nothing about this variation. Because of this, demographers often find it useful to use age-specific death rates. The age-specific death rate at age *x* years is defined as,

$$\begin{matrix}\text{age- specific death rate at} \\ \text{age x years}\end{matrix} = \frac{\text{number of deaths of people aged x years}}{\text{population aged x years}},$$

in a given calendar year. When we refer to 'age *x* years', we mean 'aged *x* last birthday'. The denominator, as before, is the mid-year population.

Denoting the age-specific death rate at age *x* years last birthday by the symbol mx, the number of deaths of people aged *x* years last birthday by θ_x, and the population aged *x* years last birthday by P_x, we can write,

$$m_x = \frac{\theta_x}{p_x},$$

or, if preferred,

$$m_x = \frac{\theta_x}{p_x} \times 1000.$$

Note that the subscripts x denote years of age, not calendar years.

Age-specific death rates can be calculated for single years of age, or for age groups, such as 5-9 years last birthday, 10-14 years last birthday, and so on. Because mortality is also known to vary by sex, age-specific death rates are usually calculated separately for males and females. When age-specific rates are calculated for age groups, a special notation is used to denote the precise age group under consideration. The symbol ${}_n\theta_x$ denotes the number of deaths to people between the exact ages x and $x+n$ years. The symbol ${}_nP_x$ is used to denote the mid-year population of people between the exact ages x and $x+n$ years, and the symbol ${}_nm_x$ denotes the age-specific death rate between exact ages x and $x+n$ years. Thus, for example, the age-specific death rate at ages 5-9 years last birthday, ${}_5m_5$, is calculated using the formula,

$${}_5m_5 = \frac{{}_5\theta_5}{{}_5p_5}.$$

To take an example, the male population aged 35-44 years last birthday in England and Wales on 30 June 1995 is estimated to have been 3 333 000. The number of deaths reported in England and Wales of males in this age group during the calendar year 1995 was 5860. The age-specific death rate in 1995 for males aged 35-44 years last birthday was, therefore, 5860/3 333 000, or 0.00176. Multiplying this by 1000 gives a rate of 1.76 per thousand.

There is one (and only one) age group for which a different method of calculating agespecific death rates is employed. This is the age group 'under 1 year', or '0 last birthday'. For this age group, the denominator is taken to be the number of live births in the calendar year in question, rather than the mid-year population aged under 1 year. For example, in England and Wales in 1995 there were 648 100 live births, and 3970 deaths to infants under 1 year. The infant mortality rate is therefore equal to 3970/648 100, which is 0.00613 or 6.13 per thousand births. Notice that this rate refers to both sexes. To measure infant mortality, unlike that of other age groups, demographers quite often use a rate referring to both sexes combined.

Two Types of Mortality Rate

1. Someone who dies in the relevant year, but before 30 June, will not be alive on that date, and will not be included in the mid-year population, yet that person's death will be included in the numerator.
2. Consider someone whose birthday is on 7 September, and who dies on 9 October. Suppose this person is aged x years last birthday on 30 June. Then when he dies he will be aged $x + 1$ years last birthday. He will be included in the denominator of the agespecific rate at age x last birthday, but his death will be included in the numerator of the age-specific rate at age $x + 1$ last birthday.

A similar problem affects the calculation of age-specific death rates for infants aged under l year when the number of births during the entire year is the denominator. Consider an infant who died on 31 March 1997 aged nine months. This child was born on 30 June 1996. His/her death is included in the numerator for the age-specific death rate at age 0 last birthday in 1997, but his/her birth is included in the denominator for the age-specific death rate at age 0 last birthday in 1996.

What can be done about this? Can rates be obtained in which the numerator and denominator correspond exactly? Yes, they can, but they require additional data. What we really need is to know the exact period of exposure at each age during the given year for each person at risk of dying. Thus, for example, the person whose $(x + l)$th birthday was on 7 September, and who died on 9 October, would be regarded as contributing 250/365 of a year's exposure during that year at age x last birthday (since there are 250 days between 1 January and 7 September), and 32/365 of a year's exposure during that year at age $x + 1$ last birthday (since there are 32 days between 7 September and 9 October). Summing these fractions of a year over the whole population under investigation for each year of age, and using the result in the denominator, would give a rate in which the numerator and denominator corresponded exactly.

In practice, such detailed information is not usually available except from special (and expensive) investigations designed to elicit it. Therefore demographers rely on mid-year populations as an *approximation* to the correct exposed-to-risk. The approximation is usually quite close in large populations.

There is, however, another approach to measuring mortality rates, which does not lead to violations of the principle of correspondence (at least at the 'person level'). In this second approach, what is done is to calculate the number of people who have their *x*th birthday during a given period, and then follow them up until either they celebrate their next birthday, or they die (whichever happens first). Dividing the number who die by the original number having their *x*th birthday gives us an age-specific death rate at age *x*.

This kind of age-specific death rate is often called a *q-type rate,* and is given the symbol *qx,* to distinguish it from the first kind of age-specific rate, *mx,* which is called an *m-type rate*. Demographers also use the terms *initial rates* for q-type rates and *central rates* for m-type rates. This is because in q-type rates the exposed-to-risk is defined at the start, or initiation, of the year of age under investigation (that is, when the members of the exposed-to-risk celebrate their *x*th birthday), whereas in m-type rates the exposed-to-risk is an estimate of the number of persons aged *x* last birthday at the time the events took place. The average age of these persons is $x+\frac{1}{2}$ years: they are half-way through (in the 'centre' of) the year of age in question.

Strictly speaking, q-type rates do not lead to an exposed-to-risk which is exactly right. Those people who die between exact ages *x* and *x* + 1 years are actually only 'at risk' of dying for the period between their *x*th birthday and the point at which they die (since once they have died, they are no longer at risk). The q-type rate assumes that such people are at risk for the entire year between exact ages *x* and *x* + 1. Thus q-type rates over-estimate the length of time exposed to risk for every person who dies. Nevertheless, they do get the right number of people in the denominator, which *m*-type rates calculated using mid-year populations cannot be relied upon to do. That is what is meant by saying that q-type rates maintain the principle of correspondence at the 'person level'.

Difference between the Two Types of Rate

The two types of mortality rate are examples of two quite different approaches to measuring the components of population change. One approach calculates rates based on a specific calendar time period (m-type rates). This is known as the period approach. The other approach calculates rates based on the experience of a specific group of people born during a specific calendar period (q-type rates). Since q-type rates are based on a group of people who celebrate their xth birthday during a given period, it follows that they must all have been born during a period of the same length x years earlier. Such a group of people is known as a birth cohort, and this approach is called the cohort approach.

Lexis Chart

The difference between the period and cohort approaches, and hence between m-type and q type rates, may be illustrated using a diagram called a Lexis chart. A Lexis chart has a vertical axis which represents age, and a horizontal axis which represents calendar time. Since people get older as time goes on, the life of any person can be represented on a Lexis chart by a diagonal line running from the horizontal axis until a point which corresponds to the person's age at death measured on the vertical axis in the figure.

On a Lexis chart, the population alive and aged x last birthday, Px at a particular point in time is represented by a vertical line. In Figure 2.2, the line AB represents those alive aged 30 years last birthday on 30 June 1997. Vertical lines like line AB thus represent the denominators of m-type mortality rates.

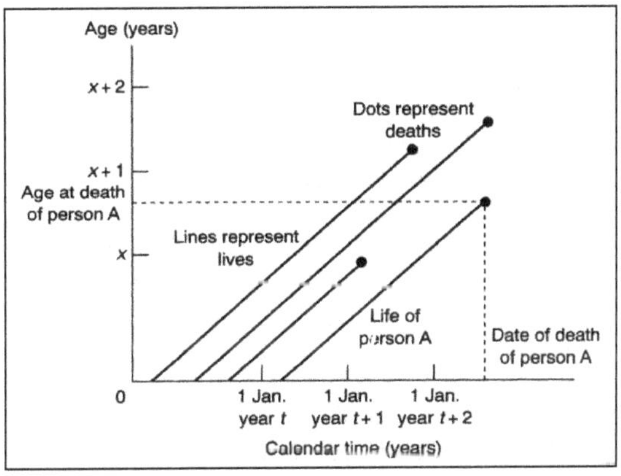

Figure: Principle behind the Lexis chart. Individual lives are represented by diagonal lines running from the bottom left towards the top right of the chart. As individuals grow older, they move up their 'life lines'. The coordinates of the upper end of each line denote the time of death and the age of the person when he/she died.

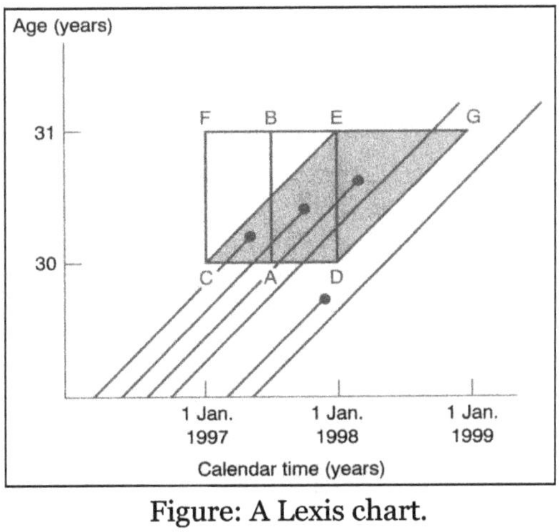

Figure: A Lexis chart.

A set of people who celebrate their xth birthday during a particular time period (which means that they were all born during a particular time period and thus constitute a birth cohort) are represented by a horizontal line. In figure, the horizontal line CD represents all the people who celebrated their 30th birthday during the year 1997 (and who were, there fore, all born during 1967). Horizontal lines like line CD represent birth cohorts, or the denominators of q-type mortality rates .

The deaths of people aged x last birthday on their date of death, and who died in a particular time period, are represented by squares. The square CDEF in figure below represents all the people who died during the calendar year 1997 who were aged 30 years last birthday when they died. Squares like this represent the numerators of m-type mortality rates.

The deaths of people aged x last birthday on their date of death, and who all celebrated their xth birthday during a particular time period, are represented by parallelograms. In figure, the shaded parallelogram CDGE represents all the people who died aged 30 years last birthday when they died, and who celebrated their 30th birthday during the calendar year 1997. Parallelograms like this represent the numerators of q-type mortality rates.

Relationship between the Two Types of Mortality Rate

In principle, there is no necessary relationship between the two types of mortality rate. However, by making a number of assumptions, a theoretical relationship can be derived. Since the assumptions are not too unreasonable, the theoretical relationship works quite well in most practical situations.

Consider the Lexis chart in figure. The deaths representing the numerator of the m type mortality rate at age x last birthday in calendar year t are in the square PQRS. Suppose that there are θx of these deaths. The deaths representing the numerator of the q-type mortality rate at age x which most closely overlaps with the numerator of the m-type rate for year t are in the shaded parallelogram TMWN.

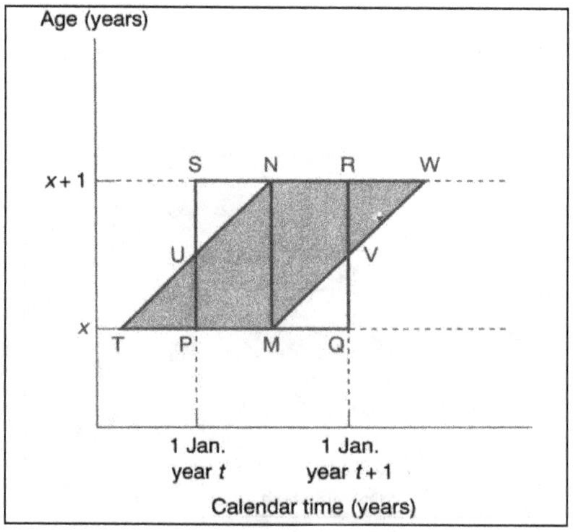

Figure: Lexis chart illustrating the relationship between the two types of mortality rate.

We now make two assumptions. We assume that:

1. Mortality only varies with age, and not with calendar time;
2. Deaths are evenly distributed across each single year of age.

These two assumptions are made repeatedly in the demographic analysis of mortality. The assumption that mortality only varies with age is easy enough to understand, but the second assumption may need clarification. The assumption that deaths are evenly distributed across each year of age implies that, say, the average age at death of those dying between their 75th and 76th birthdays is 75 years 6 months. It implies that the number of deaths between 75 years exactly and 75 years 6 months is the same as the number of deaths between ages 75 years 6 months and 76 years exactly.

Once we have made these two assumptions, then, on the Lexis chart, within any horizontal band representing the ages between x and $x+1$, deaths are evenly distributed.

This means that the number of deaths is proportional to the area of any portion of the chart. Two sections of the chart which both lie within the horizontal band representing ages between x and $x+1$ exactly and which have the same area will represent the same number of deaths.

Since the area of the parallelogram TMWN is equal to the area of the square PQRS, then the number of deaths represented by the shaded parallelogram TMWN will also be θ_x.

The m-type mortality rate, m_x, is then given by,

$$m_x = \frac{\theta_x}{\text{population represented by the vertical line MN}},$$

and the q -type mortality rate, q_x, is given by,

$$q_x = \frac{\theta_x}{\text{population represented by the horizontal line TM}}.$$

But, using the assumptions above, the number of deaths in the triangle TMN must be $\frac{1}{2}\theta_x$, and, since all the lives which cross the line TM also cross the line MN, we have

$$q_x = \frac{\theta_x}{\text{population represented by the vertical line MN} + \frac{1}{2}\theta_x,}.$$

But, from first equation above,

$$\text{population represented by the vertical line MN} = \frac{\theta_x}{m_x}.$$

Thus, substituting the above equation into the equation mentioned before the above one, we have,

$$q_x = \frac{\theta_x}{\theta_x / m_x + \frac{1}{2}\theta_x},$$

and the θ_x cancel to leave,

$$q_x = \frac{1}{1/m_x + \frac{1}{2}} = \frac{m_x}{1 + \frac{1}{2}m_x},$$

or, as it is often written,

$$q_x = \frac{2m_x}{2 + m_x}.$$

This result is dependent upon the two assumptions we have made. In many practical situations, however, the approximation is satisfactory.

There are certain age groups, though, in which the assumption that deaths are evenly distributed is not valid. This is particularly true of the first year of life. Most deaths to infants during the first year of life take place during the first few weeks of that year. Indeed, in low-mortality populations, it is usual for more than half of all the deaths to infants under the age of 1 year to occur during the first month of life. Deaths to infants during the first four weeks of life are known as neonatal deaths. Neonatal deaths may be measured using the neonatal death rate, defined as,

$$\text{neonatal death rate} = \frac{\begin{array}{c}\text{number of deaths in a given year to}\\ \text{infants aged 28 days or under}\end{array}}{\text{number of births in the given year}}.$$

For example, in the United Kingdom in 1995, there were 732000 live births, and 3070 neonatal deaths. The neonatal death rate was therefore equal to 3070/732000, which is 0.0042, or 4.2 per thousand births.

Advantages and Disadvantages of the Two Types of Mortality Rate

The two types of mortality rate have their advantages and disadvantages. Generally speaking, *m*-type rates have the advantage of being straightforward to calculate from routinely available data. Their disadvantage is that they do not reflect the experience of 'real' people, and, if calculated using mid-year populations, violate the principle of correspondence. The advantages and disadvantages of *q*-type rates are in a sense 'mirror-images' of those of *m*-type rates.

Type	Advantages	Disadvantages
m-type	Data are readily available Easy to calculate Can be applied to a specific calendar time period	Violate the principle of correspondence if based on mid-year populations Do not reflect the experience of a real group of people
q-type	Reflect the experience of real people Do not violate the principle of correspondence (at least at the 'person level')	Do not apply to a particular calendar year Data are not readily available Awkward to calculate

References

- Human-Population: biologyreference.com, Retrieved 31 March 2020
- Population-policy, encyclopedias-almanacs-transcripts-and-maps, social-sciences: encyclopedia.com, Retrieved 14 June 2020
- Population-geography-overview-1435468: thoughtco.com, Retrieved 20 June 2020
- Mortality-demography, science: britannica.com, Retrieved 20 April 2020
- Population-geography-roots-definition-nature-and-subject-matter-43085: your article library.com, Retrieved 15 May 2020
- Population-human, encyclopedias-almanacs-transcripts-and-maps, science: encyclopedia.com, Retrieved 22 June 2020

6

Understanding Demographic Models

The representation of demographic processes through a mathematical function or a set of functions connecting two or more measurable demographic variables is termed as a demographic model. It is used for the simplification of data. This chapter has been carefully written to provide an easy understanding of the varied facets of demographic models.

Demographic Models

Models are mathematical and/or statistical expressions of relationships between variables describing some major, chosen aspects of a phenomenon. In population studies they are often used to explore relationships between different components of demographic systems, and to explore relationship between demographic variables and socio-economic and cultural variables. The relationships expressed in mathematical form of any kind are called mathematical models The Exponential growth model of population (such as P t = P o Exp (r.t)) is an example of a simple mathematical model which states that the population growth follows the exponential model. When the relationships are developed using statistical methods such as regression analysis, they are called statistical models. They contain an element of uncertainty. Statistical models have been developed for studying the growth of population as a stochastic process, i.e., a process which can be described only in terms of probability distributions (Kendall , 1949). Thus statistical models are closer to reality than mathematical models. They are particularly suited for the studies of random errors. Models using probability distributions, Monte Carlo computer simulation methods (because of their reliance on repeated computation of random or pseudo-random numbers) , regression analysis, discriminant analysis, factor analysis etc. are all statistical models.

History of Mathematical Models

For long, scholars from all fields have looked for ways of representing social and scientific phenomena in the form of abstract laws, principles and mathematical forms. Population studies is no exception to this. One of the earliest and most forceful arguments in favour of modelling was, however, given by Keyfitz. In the first volume of *Population and Development Review* Keyfitz (1975) published, in an article entitled "How do we know the facts of demography?", he argued that the empirical relationships based on regression analysis can be misleading as they depend heavily on the cases for which data are available. Also, without modelling questions regarding cause and effect, multiple causation, and nature of relationships cannot be answered. In this article, he showed that the empirical relationship between percent aged 65 and over and growth rate of population was dependent on: (a) number of countries for which data are available; and (b) homogeneity among the countries. In such situations analytical and mathematical models can provide a better understanding of the relationship.

The most important of all the demographic models is the stable population model (Smith and Keyfitz, 1977; Keyfitz, 2005). It says that if individuals are born at a constant rate of 1 person per unit of time, and the survival probability of a person aged x is p(x), then at any time the expected size of the population is given by,

$$E[X] = \int_0^\alpha p(t)dt$$

where E[X] refers to the expected size of population at age x, p(t) refers to chance of survival from birth to age t, and a refers to the upper limit of the age distribution.

It may be noted that the stable population model is a deterministic model and results in a stationary population in which the number of persons at any age (x) does not change with time. Further, suppose the individuals are born at rate $Be^{\rho t}$ where ρ is a constant and t is time. Then the size of population at time t, or,

$$V(t) = Be^{at} \int_0^\alpha e^{-px} p(x)dx$$

This population grows or declines at rate ρ, and its age distribution, i.e., proportion surviving to age x is proportional to $e^{-px}p(x)$. Using the property of stable populations that age distributions of two stable populations never cross each other census growth rate and proportion of population up to a certain age (normally 35 years) were used to estimate birth and death rates for those populations which lacked reliable and complete data on them. Development of model populations, showing probability of surviving from birth to different ages at different levels of life expectancy, for different regions of the world helped the demographers working on the populations of developing countries immensely (UN, 1983). Subsequently, the stable population theory was modified to accommodate for changes in mortality or fertility or both. Nath and Kalita (1989) developed a two-sex quasi-stable population in the presence of immigration.

Stages in Modeling

Modelling requires through review of what is known and what is not known. Then the process to be modelled is analysed and represented in the form of mathematical equations, statistical relationships and/or probability distributions. The equations would contain some unknown characteristics of the population to be modelled. They are estimated using an appropriate method of fitting. This is followed by testing whether the model reproduced the old data accurately. If it did, then the model can be used for predictions or theorising. Thus modelling is done in four steps:

- specification of model;
- estimation of parameters;

- validation;
- forecasting.

Specification

Specification refers to the process of expressing relationships between dependent and independent variables in the form of mathematical or statistical equations. These relationships contain variables and some constants. The constants of the equations are called parameters. Specification requires a sound understanding of the field of study and a thorough review of works done in the field, understanding of assumptions on which past models were based, how sound they are, and the researcher's judgement. Suppose in a particular field, such as the study of waiting time in fertility, past researchers have used mostly Poisson distribution and the researcher thinks that the population being heterogeneous, a mixture of Poisson distributions will fit better. Then the researcher has to specify the waiting time as a linear combination of two Poisson distributions containing different parameters.

Estimation of Parameeters

The next step in model building is to estimate the parameters of the model. For this purpose, demographers have to use a proper estimation technique. For example, if a dependent variable is expressed as a linear combination of a number of independent variables, it contains certain parameters which may be estimated using the least square method or method of maximum likelihood. Thus the researcher has to use one method from a number of available methods of estimation given for a particular context. The researcher must know the merits and demerits of all the methods of estimation and choose on the basis of knowledge and logic.

Validation

Validation refers to examination of whether the model predicts well. In other words, validation is an examination of how close the values of dependent variable are to the values of the variable which are estimated using the model and available data on independent variables. In regression analysis R-square is used as a measure of goodness of model and a high value of R-square suggests that the model can be used for prediction and explanation.

Forecasting

Once the model is ready, it can be used for forecasting. It must be noted that an advanced and complex model is not a guarantee of good forecasts. Simple models do sometimes result in better prediction than complex models. The accuracy of forecasts depends on a number of factors: (a) model specification; (b) accuracy of data; (c) estimation procedure: (d) stability of the system; and the (d) length of the projection period. Long term forecasts cannot be as good as short term forecasts because the longer projection period is, the greater is the likelihood of changes in the system dynamics (i.e., relationships between variables).

Validity (Statistics)

Validity is the extent to which a concept, conclusion or measurement is well-founded and corresponds accurately to the real world. The word "valid" is derived from the Latin validus, meaning strong. The validity of a measurement tool (for example, a test in education) is considered to be the degree to which the tool measures what it claims to measure; in this case, the validity is an equivalent to accuracy.

In psychometrics, validity has a particular application known as test validity: "the degree to which evidence and theory support the interpretations of test scores" ("as entailed by proposed uses of tests").

It is generally accepted that the concept of scientific validity addresses the nature of reality and as such is an epistemological and philosophical issue as well as a question of measurement. The use of the term in logic is narrower, relating to the truth of inferences made from premises.

Validity is important because it can help determine what types of tests to use, and help to make sure researchers are using methods that are not only ethical, and cost-effective, but also a method that truly measures the idea or construct in question.

Test Validity

Validity (Accuracy)

Validity of an assessment is the degree to which it measures what it is supposed to measure. This is not the same as reliability, which is the extent to which a measurement gives result that are very consistent. Within validity, the measurement does not always have to be similar, as it does in reliability. However, just because a measure is reliable, it is not necessarily valid (and vice versa). Validity is also dependent on the measurement measuring what it was designed to measure, and not something else instead. Validity (similar to reliability) is a relative concept; validity is not an all-or-nothing idea. There are many different types of validity.

Construct Validity

Construct validity refers to the extent to which operationalizations of a construct (e.g., practical tests developed from a theory) measure a construct as defined by a theory. It subsumes all other types of validity. For example, the extent to which a test measures intelligence is a question of construct validity. A measure of intelligence presumes, among other things, that the measure is associated with things it should be associated with (convergent validity), not associated with things it should not be associated with (discriminant validity).

Construct validity evidence involves the empirical and theoretical support for the interpretation of the construct. Such lines of evidence include statistical analyses of the internal structure of the test including the relationships between responses to different test items. They also include relationships between the test and measures of other constructs. As currently understood, construct validity is not distinct from the support for the substantive theory of the construct that the test is designed to measure. As such, experiments designed to reveal aspects of the causal role of the construct also contribute to construct validity evidence.

Content Validity

Content validity is a non-statistical type of validity that involves "the systematic examination of the test content to determine whether it covers a representative sample of the behavior domain to be measured" (Anastasi & Urbina, 1997 p. 114). For example, does an IQ questionnaire have items covering all areas of intelligence discussed in the scientific literature?

Content validity evidence involves the degree to which the content of the test matches a content domain associated with the construct. For example, a test of the ability to add two numbers should include a range of combinations of digits. A test with only one-digit numbers, or only even numbers, would not have good coverage of the content domain. Content related evidence typically involves subject matter experts (SME's) evaluating test items against the test specifications. Before going to final administration of questionnaires the researcher should consult the validity of items against each constructs or variables and accordingly modify measurement instruments on the basis of expert's opinion.

A test has content validity built into it by careful selection of which items to include (Anastasi & Urbina, 1997). Items are chosen so that they comply with the test specification which is drawn up through a thorough examination of the subject domain. Foxcroft, Paterson, le Roux & Herbst (2004, p. 49) note that by using a panel of experts to review the test specifications and the selection of items the content validity of a test can be improved. The experts will be able to review the items and comment on whether the items cover a representative sample of the behaviour domain.

Face Validity

Face validity is an estimate of whether a test appears to measure a certain criterion; it does not guarantee that the test actually measures phenomena in that domain. Measures may have high validity, but when the test does not appear to be measuring what it is, it has low face validity. Indeed, when a test is subject to faking (malingering), low face validity might make the test more valid. Considering one may get more honest answers with lower face validity, it is sometimes important to make it appear as though there is low face validity whilst administering the measures.

Face validity is very closely related to content validity. While content validity depends on a theoretical basis for assuming if a test is assessing all domains of a certain criterion (e.g. does assessing addition skills yield in a good measure for mathematical skills? To answer this you have to know, what different kinds of arithmetic skills mathematical skills include) face validity relates to whether a test appears to be a good measure or not. This judgment is made on the "face" of the test, thus it can also be judged by the amateur.

Face validity is a starting point, but should never be assumed to be probably valid for any given purpose, as the "experts" have been wrong before—the Malleus Malificarum (Hammer of Witches) had no support for its conclusions other than the self-imagined competence of two "experts" in "witchcraft detection," yet it was used as a "test" to condemn and burn at the stake tens of thousands women as "witches."

Criterion Validity

Criterion validity evidence involves the correlation between the test and a criterion variable (or variables) taken as representative of the construct. In other words, it compares the test with other measures or outcomes (the criteria) already held to be valid. For example, employee selection tests are often validated against measures of job performance (the criterion), and IQ tests are often validated against measures of academic performance (the criterion).

If the test data and criterion data are collected at the same time, this is referred to as concurrent validity evidence. If the test data are collected first in order to predict criterion data collected at a later point in time, then this is referred to as predictive validity evidence.

Concurrent Validity

Concurrent validity refers to the degree to which the operationalization correlates with other measures of the same construct that are measured at the same time. When the measure is compared to another measure of the same type, they will be related (or correlated). Returning to the selection test example, this would mean that the tests are administered to current employees and then correlated with their scores on performance reviews.

Predictive Validity

Predictive validity refers to the degree to which the operationalization can predict (or correlate with) other measures of the same construct that are measured at some time in the future. Again, with the selection test example, this would mean that the tests are administered to applicants, all applicants are hired, their performance is reviewed at a later time, and then their scores on the two measures are correlated.

This is also when measurement predicts a relationship between what is measured and something else; predicting whether or not the other thing will happen in the future. This type of validity is important from a public view standpoint; is this going to look acceptable to the public or not?

Experimental Validity

The validity of the design of experimental research studies is a fundamental part of the scientific method, and a concern of research ethics. Without a valid design, valid scientific conclusions cannot be drawn.

Statistical Conclusion Validity

Statistical conclusion validity is the degree to which conclusions about the relationship among variables based on the data are correct or 'reasonable'. This began as being solely about whether the statistical conclusion about the relationship of the variables was correct, but now there is a movement towards moving to 'reasonable' conclusions that use: quantitative, statistical, and qualitative data.

Statistical conclusion validity involves ensuring the use of adequate sampling procedures, appropriate statistical tests, and reliable measurement procedures. As this type of validity is concerned solely with the relationship that is found among variables, the relationship may be solely a correlation.

Internal Validity

Internal validity is an inductive estimate of the degree to which conclusions about *causal* relationships can be made (e.g. cause and effect), based on the measures used, the research setting, and the whole research design. Good experimental techniques, in which the effect of an independent variable on a dependent variable is studied under highly controlled conditions, usually allow for higher degrees of internal validity than, for example, single-case designs.

Eight kinds of confounding variable can interfere with internal validity (i.e. with the attempt to isolate causal relationships):

1. History, the specific events occurring between the first and second measurements in addition to the experimental variables.
2. Maturation, processes within the participants as a function of the passage of time (not specific to particular events), e.g., growing older, hungrier, more tired, and so on.
3. Testing, the effects of taking a test upon the scores of a second testing.
4. Instrumentation, changes in calibration of a measurement tool or changes in the observers or scorers may produce changes in the obtained measurements.
5. Statistical regression, operating where groups have been selected on the basis of their extreme scores.
6. Selection, biases resulting from differential selection of respondents for the comparison groups.
7. Experimental mortality, or differential loss of respondents from the comparison groups.
8. Selection-maturation interaction, etc. e.g., in multiple-group quasi-experimental designs.

External Validity

External validity concerns the extent to which the (internally valid) results of a study can be held to be true for other cases, for example to different people, places or times. In other words, it is about whether findings can be validly generalized. If the same research study was conducted in those other cases, would it get the same results?

A major factor in this is whether the study sample (e.g. the research participants) are representative of the general population along relevant dimensions. Other factors jeopardizing external validity are:

1. Reactive or interaction effect of testing, a pretest might increase the scores on a posttest
2. Interaction effects of selection biases and the experimental variable.
3. Reactive effects of experimental arrangements, which would preclude generalization about the effect of the experimental variable upon persons being exposed to it in non-experimental settings.
4. Multiple-treatment interference, where effects of earlier treatments are not erasable.

Ecological Validity

Ecological validity is the extent to which research results can be applied to real-life situations outside of research settings. This issue is closely related to external validity but covers the question of to what degree experimental findings mirror what can be observed in the real world (ecology

= the science of interaction between organism and its environment). To be ecologically valid, the methods, materials and setting of a study must approximate the real-life situation that is under investigation.

Ecological validity is partly related to the issue of experiment versus observation. Typically in science, there are two domains of research: observational (passive) and experimental (active). The purpose of experimental designs is to test causality, so that you can infer A causes B or B causes A. But sometimes, ethical and/or methological restrictions prevent you from conducting an experiment (e.g. how does isolation influence a child's cognitive functioning?). Then you can still do research, but it is not causal, it is correlational. You can only conclude that A occurs together with B. Both techniques have their strengths and weaknesses.

Relationship to Internal Validity

On first glance, internal and external validity seem to contradict each other – to get an experimental design you have to control for all interfering variables. That is why you often conduct your experiment in a laboratory setting. While gaining internal validity (excluding interfering variables by keeping them constant) you lose ecological or external validity because you establish an artificial laboratory setting. On the other hand, with observational research you can not control for interfering variables (low internal validity) but you can measure in the natural (ecological) environment, at the place where behavior normally occurs. However, in doing so, you sacrifice internal validity.

The apparent contradiction of internal validity and external validity is, however, only superficial. The question of whether results from a particular study generalize to other people, places or times arises only when one follows an inductivist research strategy. If the goal of a study is to deductively test a theory, one is only concerned with factors which might undermine the rigor of the study, i.e. threats to internal validity.

Diagnostic Validity

In psychiatry there is a particular issue with assessing the validity of the diagnostic categories themselves. In this context:

- content validity may refer to symptoms and diagnostic criteria;
- concurrent validity may be defined by various correlates or markers, and perhaps also treatment response;
- predictive validity may refer mainly to diagnostic stability over time;
- discriminant validity may involve delimitation from other disorders.

Robins and Guze proposed in 1970 what were to become influential formal criteria for establishing the validity of psychiatric diagnoses. They listed five criteria:

- distinct clinical description (including symptom profiles, demographic characteristics, and typical precipitants);
- laboratory studies (including psychological tests, radiology and postmortem findings);

- delimitation from other disorders (by means of exclusion criteria);
- follow-up studies showing a characteristic course (including evidence of diagnostic stability);
- family studies showing familial clustering.

These were incorporated into the Feighner Criteria and Research Diagnostic Criteria that have since formed the basis of the DSM and ICD classification systems.

Kendler in 1980 distinguished between:

- antecedent validators (familial aggregation, premorbid personality, and precipitating factors);
- concurrent validators (including psychological tests);
- predictive validators (diagnostic consistency over time, rates of relapse and recovery, and response to treatment).

Nancy Andreasen (1995) listed several additional validators – molecular genetics and molecular biology, neurochemistry, neuroanatomy, neurophysiology, and cognitive neuroscience – that are all potentially capable of linking symptoms and diagnoses to their neural substrates.

Kendell and Jablinsky (2003) emphasized the importance of distinguishing between validity and utility, and argued that diagnostic categories defined by their syndromes should be regarded as valid only if they have been shown to be discrete entities with natural boundaries that separate them from other disorders.

Kendler (2006) emphasized that to be useful, a validating criterion must be sensitive enough to validate most syndromes that are true disorders, while also being specific enough to invalidate most syndromes that are not true disorders. On this basis, he argues that a Robins and Guze criterion of "runs in the family" is inadequately specific because most human psychological and physical traits would qualify - for example, an arbitrary syndrome comprising a mixture of "height over 6 ft, red hair, and a large nose" will be found to "run in families" and be "hereditary", but this should not be considered evidence that it is a disorder. Kendler has further suggested that "essentialist" gene models of psychiatric disorders, and the hope that we will be able to validate categorical psychiatric diagnoses by "carving nature at its joints" solely as a result of gene discovery, are implausible.

In the United States Federal Court System validity and reliability of evidence is evaluated using the Daubert Standard: see Daubert v. Merrell Dow Pharmaceuticals. Perri and Lichtenwald (2010) provide a starting point for a discussion about a wide range of reliability and validity topics in their analysis of a wrongful murder conviction.

Reliability (Statistics)

Reliability in statistics and psychometrics is the overall consistency of a measure. A measure is said to have a high reliability if it produces similar results under consistent conditions. "It is the characteristic of a set of test scores that relates to the amount of random error from the measurement process that might be embedded in the scores. Scores that are highly reliable are accurate, reproducible, and consistent from one testing occasion to another. That is, if the testing process were repeated with a group of test takers, essentially the same results would be obtained. Various kinds of reliability coefficients, with values ranging between 0.00 (much error) and 1.00 (no error), are usually used to indicate the amount of error in the scores." For example, measurements of people's height and weight are often extremely reliable.

Types

There are several general classes of reliability estimates:

- Inter-rater reliability assesses the degree of agreement between two or more raters in their appraisals.
- Test-retest reliability assesses the degree to which test scores are consistent from one test administration to the next. Measurements are gathered from a single rater who uses the same methods or instruments and the same testing conditions. This includes intra-rater reliability.
- Inter-method reliability assesses the degree to which test scores are consistent when there is a variation in the methods or instruments used. This allows inter-rater reliability to be ruled out. When dealing with forms, it may be termed parallel-forms reliability.
- Internal consistency reliability, assesses the consistency of results across items within a test.

Difference from Validity

Reliability does not imply validity. That is, a reliable measure that is measuring something consistently is not necessarily measuring what you want to be measuring. For example, while there are many reliable tests of specific abilities, not all of them would be valid for predicting, say, job performance.

While reliability does not imply validity, reliability does place a limit on the overall validity of a test. A test that is not perfectly reliable cannot be perfectly valid, either as a means of measuring attributes of a person or as a means of predicting scores on a criterion. While a reliable test may provide useful valid information, a test that is not reliable cannot possibly be valid.

For example, if a set of weighing scales consistently measured the weight of an object as 500 grams over the true weight, then the scale would be very reliable, but it would not be valid (as the returned weight is not the true weight). For the scale to be valid, it should return the true weight of an object. This example demonstrates that a perfectly reliable measure is not necessarily valid, but that a valid measure necessarily must be reliable.

General Model

In practice, testing measures are never perfectly consistent. Theories of test reliability have been developed to estimate the effects of inconsistency on the accuracy of measurement. The basic starting point for almost all theories of test reliability is the idea that test scores reflect the influence of two sorts of factors.

1. Factors that contribute to consistency: stable characteristics of the individual or the attribute that one is trying to measure.

2. Factors that contribute to inconsistency: features of the individual or the situation that can affect test scores but have nothing to do with the attribute being measured.

These factors include:

- Temporary but general characteristics of the individual: health, fatigue, motivation, emotional strain.
- Temporary and specific characteristics of individual: comprehension of the specific test task, specific tricks or techniques of dealing with the particular test materials, fluctuations of memory, attention or accuracy.
- Aspects of the testing situation: freedom from distractions, clarity of instructions, interaction of personality, sex, or race of examiner.
- Chance factors: luck in selection of answers by sheer guessing, momentary distractions.

The goal of estimating reliability is to determine how much of the variability in test scores is due to errors in measurement and how much is due to variability in true scores.

A true score is the replicable feature of the concept being measured. It is the part of the observed score that would recur across different measurement occasions in the absence of error.

Errors of measurement are composed of both random error and systematic error. It represents the discrepancies between scores obtained on tests and the corresponding true scores.

This conceptual breakdown is typically represented by the simple equation:

Observed test score = true score + errors of measurement

Classical Test Theory

The goal of reliability theory is to estimate errors in measurement and to suggest ways of improving tests so that errors are minimized.

The central assumption of reliability theory is that measurement errors are essentially random. This does not mean that errors arise from random processes. For any individual, an error in measurement is not a completely random event. However, across a large number of individuals, the causes of measurement error are assumed to be so varied that measure errors act as random variables.

If errors have the essential characteristics of random variables, then it is reasonable to assume that errors are equally likely to be positive or negative, and that they are not correlated with true scores or with errors on other tests.

It is assumed that:

1. Mean error of measurement = 0.
2. True scores and errors are uncorrelated.

3. Errors on different measures are uncorrelated.

Reliability theory shows that the variance of obtained scores is simply the sum of the variance of true scores plus the variance of errors of measurement.

$$\sigma_X^2 = \sigma_T^2 + \sigma_E^2$$

This equation suggests that test scores vary as the result of two factors:

1. Variability in true scores.
2. Variability due to errors of measurement.

The reliability coefficient $\rho_{xx'}$ provides an index of the relative influence of true and error scores on attained test scores. In its general form, the reliability coefficient is defined as the ratio of *true score* variance to the total variance of test scores. Or, equivalently, one minus the ratio of the variation of the *error score* and the variation of the *observed score*:

$$\rho_{xx'} = \frac{\sigma_T^2}{\sigma_X^2} = 1 - \frac{\sigma_E^2}{\sigma_X^2}$$

Unfortunately, there is no way to directly observe or calculate the true score, so a variety of methods are used to estimate the reliability of a test.

Some examples of the methods to estimate reliability include test-retest reliability, internal consistency reliability, and *parallel-test reliability*. Each method comes at the problem of figuring out the source of error in the test somewhat differently.

Item Response Theory

It was well-known to classical test theorists that measurement precision is not uniform across the scale of measurement. Tests tend to distinguish better for test-takers with moderate trait levels and worse among high- and low-scoring test-takers. Item response theory extends the concept of reliability from a single index to a function called the *information function*. The IRT information function is the inverse of the conditional observed score standard error at any given test score.

Estimation

The goal of estimating reliability is to determine how much of the variability in test scores is due to errors in measurement and how much is due to variability in true scores.

Four practical strategies have been developed that provide workable methods of estimating test reliability.

1. Test-retest reliability method: directly assesses the degree to which test scores are consistent from one test administration to the next.

It involves:

- Administering a test to a group of individuals.

- Re-administering the same test to the same group at some later time.
- Correlating the first set of scores with the second.

The correlation between scores on the first test and the scores on the retest is used to estimate the reliability of the test using the Pearson product-moment correlation coefficient.

2. Parallel-forms method:

The key to this method is the development of alternate test forms that are equivalent in terms of content, response processes and statistical characteristics. For example, alternate forms exist for several tests of general intelligence, and these tests are generally seen equivalent.

With the parallel test model it is possible to develop two forms of a test that are equivalent in the sense that a person's true score on form A would be identical to their true score on form B. If both forms of the test were administered to a number of people, differences between scores on form A and form B may be due to errors in measurement only.

It involves:

- Administering one form of the test to a group of individuals.
- At some later time, administering an alternate form of the same test to the same group of people.
- Correlating scores on form A with scores on form B.

The correlation between scores on the two alternate forms is used to estimate the reliability of the test.

This method provides a partial solution to many of the problems inherent in the test-retest reliability method. For example, since the two forms of the test are different, carryover effect is less of a problem. Reactivity effects are also partially controlled; although taking the first test may change responses to the second test. However, it is reasonable to assume that the effect will not be as strong with alternate forms of the test as with two administrations of the same test.

However, this technique has its disadvantages:

- It may very difficult to create several alternate forms of a test.
- It may also be difficult if not impossible to guarantee that two alternate forms of a test are parallel measures.

3. Split-half method:

This method treats the two halves of a measure as alternate forms. It provides a simple solution to the problem that the parallel-forms method faces: the difficulty in developing alternate forms.

It involves:

- Administering a test to a group of individuals.

- Splitting the test in half.
- Correlating scores on one half of the test with scores on the other half of the test.

The correlation between these two split halves is used in estimating the reliability of the test. This halves reliability estimate is then stepped up to the full test length using the Spearman–Brown prediction formula.

There are several ways of splitting a test to estimate reliability. For example, a 40-item vocabulary test could be split into two subtests, the first one made up of items 1 through 20 and the second made up of items 21 through 40. However, the responses from the first half may be systematically different from responses in the second half due to an increase in item difficulty and fatigue.

In splitting a test, the two halves would need to be as similar as possible, both in terms of their content and in terms of the probable state of the respondent. The simplest method is to adopt an odd-even split, in which the odd-numbered items form one half of the test and the even-numbered items form the other. This arrangement guarantees that each half will contain an equal number of items from the beginning, middle, and end of the original test.

4. Internal consistency: assesses the consistency of results across items within a test. The most common internal consistency measure is Cronbach's alpha, which is usually interpreted as the mean of all possible split-half coefficients. Cronbach's alpha is a generalization of an earlier form of estimating internal consistency, Kuder–Richardson Formula 20. Although the most commonly used, there are some misconceptions regarding Cronbach's alpha.

These measures of reliability differ in their sensitivity to different sources of error and so need not be equal. Also, reliability is a property of the *scores of a measure* rather than the measure itself and are thus said to be *sample dependent.* Reliability estimates from one sample might differ from those of a second sample (beyond what might be expected due to sampling variations) if the second sample is drawn from a different population because the true variability is different in this second population. (This is true of measures of all types—yardsticks might measure houses well yet have poor reliability when used to measure the lengths of insects.)

Reliability may be improved by clarity of expression (for written assessments), lengthening the measure, and other informal means. However, formal psychometric analysis, called item analysis, is considered the most effective way to increase reliability. This analysis consists of computation of item difficulties and item discrimination indices, the latter index involving computation of correlations between the items and sum of the item scores of the entire test. If items that are too difficult, too easy, and/or have near-zero or negative discrimination are replaced with better items, the reliability of the measure will increase.

- $R(t) = 1 - F(t)$.
- $R(t) = \exp(-\lambda\ )$. (where λ is the failure rate).

Replication (Statistics)

In engineering, science, and statistics, replication is the repetition of an experimental condition so that the variability associated with the phenomenon can be estimated. ASTM, in standard E1847, defines replication as "the repetition of the set of all the treatment combinations to be compared in an experiment. Each of the repetitions is called a replicate."

Replication is not the same as repeated measurements of the same item: they are dealt with differently in statistical experimental design and data analysis.

For proper sampling, a process or batch of products should be in reasonable statistical control; inherent random variation is present but variation due to assignable (special) causes is not. Evaluation or testing of a single item does not allow for item-to-item variation and may not represent the batch or process. Replication is needed to account for this variation among items and treatments.

Example

As an example, consider a continuous process which produces items. Batches of items are then processed or treated. Finally, tests or measurements are conducted. Several options might be available to obtain ten test values. Some possibilities are:

- One finished and treated item might be measured repeatedly to obtain ten test results. Only one item was measured so there is no replication. The repeated measurements help identify observational error.
- Ten finished and treated items might be taken from a batch and each measured once. This is not full replication because the ten samples are not random and not representative of the continuous nor batch processing.
- Five items are taken from the continuous process based on sound statistical sampling. These are processed in a batch and tested twice each. This includes replication of initial samples but does not allow for batch-to-batch variation in processing. The repeated tests on each provide some measure and control of testing error.
- Five items are taken from the continuous process based on sound statistical sampling. These are processed in five different batches and tested twice each. This plan includes proper replication of initial samples and also includes batch-to-batch variation. The repeated tests on each provide some measure and control of testing error.

Each option would call for different data analysis methods and yield different conclusions.

Issues in Modeling

The major issues in population studies are as follows:

- growth of population;
- changes in composition of population;
- demographic processes such as nuptiality (i.e., marriage), fertility, mortality and migration.

Mathematical and statistical models have been used in studies of all the above issues. Among them, as compared to nuptiality, more attention has been paid to modelling of fertility and mortality.

Predicting Growth of Populations and Subpopulations

Mathematical models have been used commonly to predict the size and composition of population of a country or a geographical region. For this purpose, various functions such as linear function, geometric or exponential growth function, modified exponential function, logistic curve, Makeham curve, Gompertz curve, polynomials, hyperbolic functions and autoregressive series have been used (Misra, 1980). Among them, logistic curve has found support more than any other function on empirical grounds as well as the logic that the "population increase is proportional to the absolute population size already attained and the amount still left until the maximum, where the population becomes stationary" (UN, 1973).

Logistic Model

Among all the growth models existing in literature, logistic model is most widely used model of population growth. The model and its assumptions are discussed below.

Assuming that r is the maximum rate of increase of the population,

$$\frac{K-N}{K}$$

the fraction by which the actual population (N) remains below the maximum (K), then the increase of population per unit of time is:

$$\frac{dN}{dT} = rN\frac{K-N}{K}$$

The total population will then be:

$$N = \frac{K}{1+be^{-rt}}$$

Figure shows the various forms of logistic growth curve for different values of t and r (Weisstein, Eric W., 2009). In a typical logistic growth model, showing growth of population, N increases continuously from zero to a saturating level in a certain manner: initially when N is small, the growth rate is also small, but it continues to increase till a maximum point of growth rate is reached after which although N continues to rise the rate of growth becomes smaller and smaller, ultimately reaching zero. Is this not a realistic assumption for population growth? Almost the same thing can be said about the degree of urbanization, spread of literacy and many other things in population studies. All these processes can be modelled using logistic curve.

Logistic regression is a type of regression model in which log of odds ratio of a binary/qualitative dependent variable is expressed as a linear function of a number of qualitative as well as quantitative variables. During the last fifty years logistic regression has completely replaced the simple multiple regression analysis in sociology.

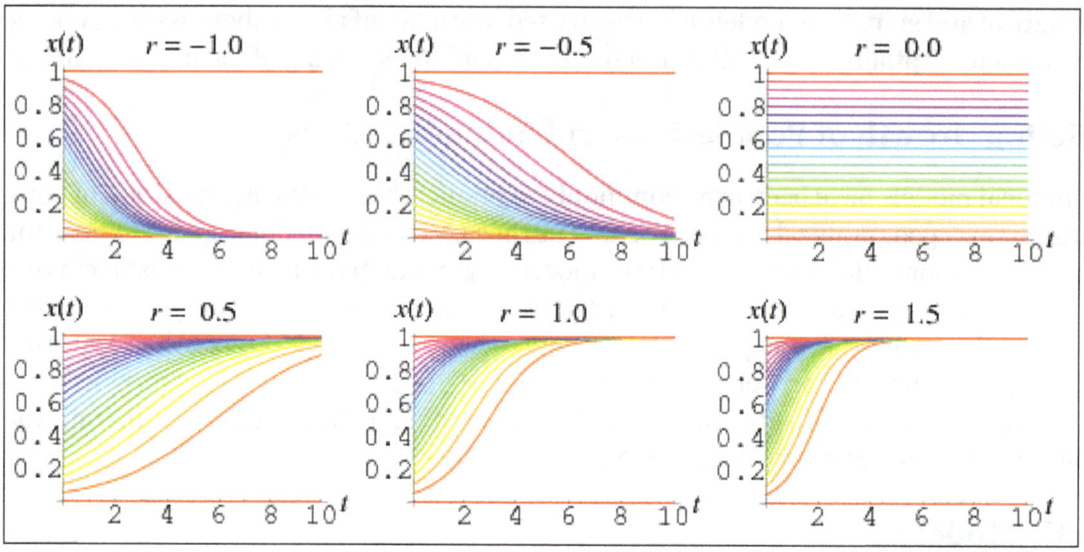

Forms of Logistic Growth Curve.

In the above equation r is also called the Malthusian parameter (being rate of maximum population growth) and K is called the carrying capacity (i.e., the maximum sustainable population). Dividing both sides by K and defining x as N/K gives the differential equation,

$$\frac{dx}{dt} = r\,x\,(1 - x)$$

Growth rate of a population following logistic curve has a definite pattern: rising from an extremely low level to a maximum level sometime and declining after that, gradually reaching zero. Logistic model has also been found to be of immense use in predicting subpopulations (Leach, 1981) because it provides a working model of the "mechanism of self correction". UN Manual XIII (UN, 1974) showed that a constant urban-rural growth difference (URGD) leads to logistic growth of the degree of urbanization (i.e., percent urban).

Subsequently, URGD method was used for projections of ratio of urban population to total population, and even other types of ratios such as ratios of populations of cities to total urban population, ratios of labour force to total population and school enrolment rates. Thus urban population can be predicted by multiplying projections of total population by projected ratios of urban population to total population. Projections of total population are obtained using component method (Smith, 1992).

Logistic Regression

In statistics, logistic regression, or logit regression, or logit model is a regression model where the dependent variable (DV) is categorical. This section covers the case of a binary dependent variable—that is, where it can take only two values, "0" and "1", which represent outcomes such as pass/fail, win/lose, alive/dead or healthy/sick. Cases where the dependent variable has more than two outcome categories may be analysed in multinomial logistic regression, or, if the multiple categories are ordered, in ordinal logistic regression. In the terminology of economics, logistic regression is an example of a qualitative response/discrete choice model.

Logistic regression was developed by statistician David Cox in 1958. The binary logistic model is used to estimate the probability of a binary response based on one or more predictor (or independent) variables (features). It allows one to say that the presence of a risk factor increases the probability of a given outcome by a specific percentage.

Fields and Example Applications

Logistic regression is used in various fields, including machine learning, most medical fields, and social sciences. For example, the Trauma and Injury Severity Score (TRISS), which is widely used to predict mortality in injured patients, was originally developed by Boyd et al. using logistic regression. Many other medical scales used to assess severity of a patient have been developed using logistic regression. Logistic regression may be used to predict whether a patient has a given disease (e.g. diabetes; coronary heart disease), based on observed characteristics of the patient (age, sex, body mass index, results of various blood tests, etc.). Another example might be to predict whether an American voter will vote Democratic or Republican, based on age, income, sex, race, state of residence, votes in previous elections, etc. The technique can also be used in engineering, especially for predicting the probability of failure of a given process, system or product. It is also used in marketing applications such as prediction of a customer's propensity to purchase a product or halt a subscription, etc. In economics it can be used to predict the likelihood of a person's choosing to be in the labor force, and a business application would be to predict the likelihood of a homeowner defaulting on a mortgage. Conditional random fields, an extension of logistic regression to sequential data, are used in natural language processing.

Example: Probability of Passing an Exam Versus Hours of Study

Suppose we wish to answer the following question:

A group of 20 students spend between 0 and 6 hours studying for an exam. How does the number of hours spent studying affect the probability that the student will pass the exam?

The reason for using logistic regression for this problem is that the dependent variable pass/fail represented by "1" and "0" are not cardinal numbers. If the problem were changed so that pass/fail was replaced with the grade 0–100 (cardinal numbers), then simple regression analysis could be used.

The table shows the number of hours each student spent studying, and whether they passed (1) or failed (0).

Hours	0.50	0.75	1.00	1.25	1.50	1.75	1.75	2.00	2.25	2.50	2.75	3.00	3.25	3.50	4.00	4.25	4.50	4.75	5.00	5.50
Pass	0	0	0	0	0	0	1	0	1	0	1	0	1	0	1	1	1	1	1	1

The graph shows the probability of passing the exam versus the number of hours studying, with the logistic regression curve fitted to the data.

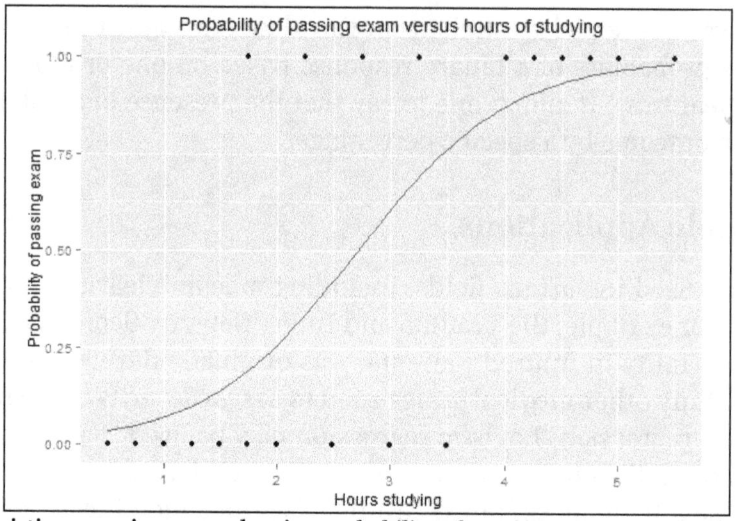

Graph of a logistic regression curve showing probability of passing an exam versus hours studying.

The logistic regression analysis gives the following output.

	Coefficient	Std.Error	z-value	P-value (Wald)
Intercept	−4.0777	1.7610	−2.316	0.0206
Hours	1.5046	0.6287	2.393	0.0167

The output indicates that hours studying is significantly associated with the probability of passing the exam ($p = 0.0167$, Wald test). The output also provides the coefficients for $\text{Intercept} = -4.0777$ and $\text{Hours} = 1.5046$. These coefficients are entered in the logistic regression equation to estimate the probability of passing the exam:

$$\text{Probability of passing exam} = \frac{1}{1 + \exp\left(-\left(1.5046 \cdot \text{Hours} - 4.0777\right)\right)}$$

For example, for a student who studies 2 hours, entering the value $\text{Hours} = 2$ in the equation gives the estimated probability of passing the exam of 0.26:

$$\text{Probability of passing exam} = \frac{1}{1 + \exp\left(-\left(1.5046 \cdot 2 - 4.0777\right)\right)} = 0.26$$

Similarly, for a student who studies 4 hours, the estimated probability of passing the exam is 0.87:

$$\text{Probability of passing exam} = \frac{1}{1 + \exp\left(-\left(1.5046 \cdot 4 - 4.0777\right)\right)} = 0.87$$

This table shows the probability of passing the exam for several values of hours studying.

Hours of study	Probability of passing exam
1	0.07
2	0.26

3	0.61
4	0.87
5	0.97

The output from the logistic regression analysis gives a p-value of $p = 0.0167$, which is based on the Wald z-score. Rather than the Wald method, the recommended method to calculate the p-value for logistic regression is the Likelihood Ratio Test (LRT), which for this data gives $p = 0.0006$.

Some Discussion

Logistic regression can be binomial, ordinal or multinomial. Binomial or binary logistic regression deals with situations in which the observed outcome for a dependent variable can have only two possible types, "0" and "1" (which may represent, for example, "dead" vs. "alive" or "win" vs. "loss"). Multinomial logistic regression deals with situations where the outcome can have three or more possible types (e.g., "disease A" vs. "disease B" vs. "disease C") that are not ordered. Ordinal logistic regression deals with dependent variables that are ordered. In binary logistic regression, the outcome is usually coded as "0" or "1", as this leads to the most straightforward interpretation. If a particular observed outcome for the dependent variable is the noteworthy possible outcome (referred to as a "success" or a "case") it is usually coded as "1" and the contrary outcome (referred to as a "failure" or a "noncase") as "0". Logistic regression is used to predict the odds of being a case based on the values of the independent variables (predictors). The odds are defined as the probability that a particular outcome is a case divided by the probability that it is a noncase.

Like other forms of regression analysis, logistic regression makes use of one or more predictor variables that may be either continuous or categorical. Unlike ordinary linear regression, however, logistic regression is used for predicting binary dependent variables (treating the dependent variable as the outcome of a Bernoulli trial) rather than a continuous outcome. Given this difference, the assumptions of linear regression are violated. In particular, the residuals cannot be normally distributed. In addition, linear regression may make nonsensical predictions for a binary dependent variable. What is needed is a way to convert a binary variable into a continuous one that can take on any real value (negative or positive). To do that logistic regression first takes the odds of the event happening for different levels of each independent variable, then takes the ratio of those odds (which is continuous but cannot be negative) and then takes the logarithm of that ratio. This is referred to as logit or log-odds) to create a continuous criterion as a transformed version of the dependent variable.

Thus the logit transformation is referred to as the *link function* in logistic regression—although the dependent variable in logistic regression is binomial, the logit is the continuous criterion upon which linear regression is conducted.

The logit of success is then fitted to the predictors using linear regression analysis. The predicted value of the logit is converted back into predicted odds via the inverse of the natural logarithm, namely the exponential function. Thus, although the observed dependent variable in logistic regression is a zero-or-one variable, the logistic regression estimates the odds, as a continuous variable, that the dependent variable is a success (a case). In some applications the odds are all that is needed. In others, a specific yes-or-no prediction is needed for whether the dependent variable is or is not a case; this categorical prediction can be based on the computed odds of a success, with predicted odds above some chosen cutoff value being translated into a prediction of a success.

Logistic Regression vs. other Approaches

Logistic regression measures the relationship between the categorical dependent variable and one or more independent variables by estimating probabilities using a logistic function, which is the cumulative logistic distribution. Thus, it treats the same set of problems as probit regression using similar techniques, with the latter using a cumulative normal distribution curve instead. Equivalently, in the latent variable interpretations of these two methods, the first assumes a standard logistic distribution of errors and the second a standard normal distribution of errors.

Logistic regression can be seen as a special case of the generalized linear model and thus analogous to linear regression. The model of logistic regression, however, is based on quite different assumptions (about the relationship between dependent and independent variables) from those of linear regression. In particular the key differences of these two models can be seen in the following two features of logistic regression. First, the conditional distribution is $y|x$ a Bernoulli distribution rather than a Gaussian distribution, because the dependent variable is binary. Second, the predicted values are probabilities and are therefore restricted to (0,1) through the logistic distribution function because logistic regression predicts the probability of particular outcomes.

Logistic regression is an alternative to Fisher's 1936 method, linear discriminant analysis. If the assumptions of linear discriminant analysis hold, the conditioning can be reversed to produce logistic regression. The converse is not true, however, because logistic regression does not require the multivariate normal assumption of discriminant analysis.

Latent Variable Interpretation

The logistic regression can be understood simply as finding the β parameters that best fit:

$$y = \begin{cases} 1 & \beta_0 + \beta_1 x + \varepsilon > 0 \\ 0 & \text{else} \end{cases}$$

where ε is an error distributed by the standard logistic distribution. (If the standard normal distribution is used instead, it is a probit regression.)

The associated latent variable is $y' = \beta_0 + \beta_1 x + \varepsilon$. The error term ε is not observed, and so the y' is also an unobservable, hence termed "latent". (The observed data are values of y and x.) Unlike ordinary regression, however, the β parameters cannot be expressed by any direct formula of the y and values in the observed data. Instead they are to be found by an iterative search process, usually implemented by a software program, that finds the maximum of a complicated "likelihood expression" that is a function of all of the observed y and x values. The estimation approach is explained below.

Logistic Function, Odds, Odds Ratio, and Logit

Definition of the Logistic Function

An explanation of logistic regression can begin with an explanation of the standard logistic function. The logistic function is useful because it can take any real input t, $(t \in \mathbb{R})$, whereas the output always takes values between zero and one and hence is interpretable as a probability. The logistic function $\sigma(t)$ is defined as follows;

$$\sigma(t) = \frac{e^t}{e^t + 1} = \frac{1}{1 + e^{-t}}$$

A graph of the logistic function on the *t*-interval (-6,6) is shown in Figure.

Let us assume that t is a linear function of a single explanatory variable x (the case where t is a *linear combination* of multiple explanatory variables is treated similarly). We can then express t as follows:

$$t = \beta_0 + \beta_1 x$$

And the logistic function can now be written as:

$$F(x) = \frac{1}{1 + e^{-(\beta_0 + \beta_1 x)}}$$

Note that $F(x)$ is interpreted as the probability of the dependent variable equaling a "success" or "case" rather than a failure or non-case. It's clear that the response variables Y_i are not identically distributed: $P(Y_i = 1 \mid X)$ differs from one data point X_i to another, though they are independent given design matrix X and shared with parameters β.

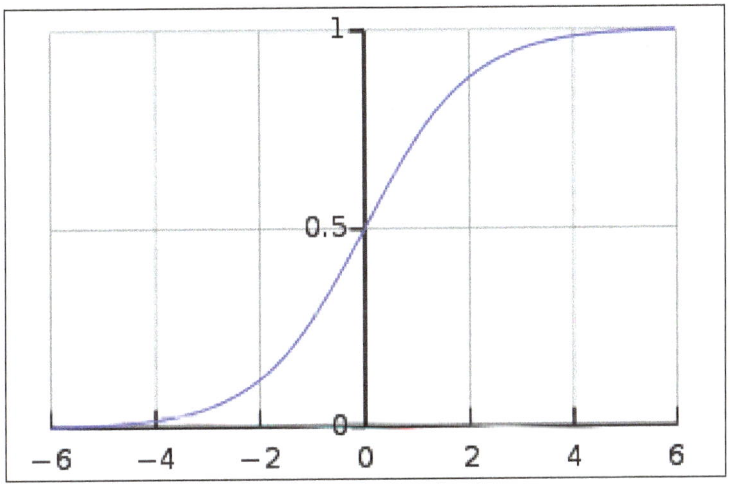

The standard logistic function $\sigma(t)$; note that $\sigma(t) \in (0,1)$ for all t.

Definition of the Inverse of the Logistic Function

We can now define the inverse of the logistic function, g, the logit (log odds):

$$g(F(x)) = \ln\left(\frac{F(x)}{1 - F(x)}\right) = \beta_0 + \beta_1 x,$$

and equivalently, after exponentiating both sides:

$$\frac{F(x)}{1 - F(x)} = e^{\beta_0 + \beta_1 x}.$$

Interpretation of These Terms

In the above equations, the terms are as follows:

- g is the logit function. The equation for $g(F(x))$ illustrates that the logit (i.e., log-odds or natural logarithm of the odds) is equivalent to the linear regression expression.
- ln denotes the natural logarithm.
- $F(x)$ is the probability that the dependent variable equals a case, given some linear combination of the predictors. The formula for $F(x)$ illustrates that the probability of the dependent variable equaling a case is equal to the value of the logistic function of the linear regression expression. This is important in that it shows that the value of the linear regression expression can vary from negative to positive infinity and yet, after transformation, the resulting expression for the probability $F(x)$ ranges between 0 and 1.
- β_0 is the intercept from the linear regression equation (the value of the criterion when the predictor is equal to zero).
- $\beta_1 x$ is the regression coefficient multiplied by some value of the predictor.
- Base e denotes the exponential function.

Definition of the Odds

The odds of the dependent variable equaling a case (given some linear combination x of the predictors) is equivalent to the exponential function of the linear regression expression. This illustrates how the logit serves as a link function between the probability and the linear regression expression. Given that the logit ranges between negative and positive infinity, it provides an adequate criterion upon which to conduct linear regression and the logit is easily converted back into the odds.

So we define odds of the dependent variable equaling a case (given some linear combination x of the predictors) as follows:

$$\text{odds} = e^{\beta_0 + \beta_1 x}.$$

The Odds Ratio

For a continuous independent variable the odds ratio can be defined as:

$$\text{OR} = \frac{\text{odds}(x+1)}{\text{odds}(x)} = \frac{\left(\frac{F(x+1)}{1-F(x+1)}\right)}{\left(\frac{F(x)}{1-F(x)}\right)} = \frac{e^{\beta_0 + \beta_1 (x+1)}}{e^{\beta_0 + \beta_1 x}} = e^{\beta_1}$$

This exponential relationship provides an interpretation for β_1: The odds multiply by e^{β_1} for every 1-unit increase in x.

For a binary independent variable the odds ratio is defined as $\frac{ad}{bc}$ where a, b, c and d are cells in a 2×2 contingency table.

Multiple Explanatory Variables

If there are multiple explanatory variables, the above expression $\beta_0 + \beta_1 x$ can be revised to $\beta_0 + \beta_1 x_1 + \beta_2 x_2 + \cdots + \beta_m x_m$. Then when this is used in the equation relating the logged odds of a success to the values of the predictors, the linear regression will be a multiple regression with m explanators; the parameters β_j for all j = 0, 1, 2, ..., m are all estimated.

Model Fitting

Estimation

Because the model can be expressed as a generalized linear model, for $0 < p < 1$, ordinary least squares can suffice, with R-squared as the measure of goodness of fit in the fitting space. When $p = 0$ or 1, more complex methods are required.

"Rule of Ten"

A widely-used rule of thumb states that logistic regression models give stable values for the explanatory variables if based on a minimum of about 10 events per explanatory variable (EPV); where *event* denotes the cases belonging to the less frequent category in the dependent variable. Thus a study designed to use k explanatory variables for an event (eg myocardial infarction) expected to occur in a proportion p of participants in the study will require a total of $10k / p$ participants. However, there is considerable debate about the reliability of this rule, which is based on simulation studies and lacks a secure theoretical underpinning. According to some authors the rule is overconservative, some circumstances; with the authors stating "If we (somewhat subjectively) regard confidence interval coverage less than 93 percent, type I error greater than 7 percent, or relative bias greater than 15 percent as problematic, our results indicate that problems are fairly frequent with 2–4 EPV, uncommon with 5–9 EPV, and still observed with 10–16 EPV. Cox models appear to be slightly more susceptible than logistic. The worst instances of each problem were not severe with 5–9 EPV and usually comparable to those with 10–16 EPV".

Maximum Likelihood Estimation

The regression coefficients are usually estimated using maximum likelihood estimation. Unlike linear regression with normally distributed residuals, it is not possible to find a closed-form expression for the coefficient values that maximize the likelihood function, so that an iterative process must be used instead; for example Newton's method. This process begins with a tentative solution, revises it slightly to see if it can be improved, and repeats this revision until no more improvement is made, at which point the process is said to have converged.

In some instances the model may not reach convergence. Nonconvergence of a model indicates that the coefficients are not meaningful because the iterative process was unable to find appropriate solutions. A failure to converge may occur for a number of reasons: having a large ratio of predictors to cases, multicollinearity, sparseness, or complete separation.

- Having a large ratio of variables to cases results in an overly conservative Wald statistic and can lead to nonconvergence.
- Multicollinearity refers to unacceptably high correlations between predictors. As multicollinearity increases, coefficients remain unbiased but standard errors increase and the likelihood of model convergence decreases. To detect multicollinearity amongst the predictors, one can conduct a linear regression analysis with the predictors of interest for the sole purpose of examining the tolerance statistic used to assess whether multicollinearity is unacceptably high.
- Sparseness in the data refers to having a large proportion of empty cells (cells with zero counts). Zero cell counts are particularly problematic with categorical predictors. With continuous predictors, the model can infer values for the zero cell counts, but this is not the case with categorical predictors. The model will not converge with zero cell counts for categorical predictors because the natural logarithm of zero is an undefined value, so that final solutions to the model cannot be reached. To remedy this problem, researchers may collapse categories in a theoretically meaningful way or add a constant to all cells.
- Another numerical problem that may lead to a lack of convergence is complete separation, which refers to the instance in which the predictors perfectly predict the criterion – all cases are accurately classified. In such instances, one should reexamine the data, as there is likely some kind of error.

Iteratively Reweighted Least Squares (IRLS)

Binary logistic regression ($y = 0$ or $y = 1$) can, for example, be calculated using *iteratively reweighted least squares* (IRLS), which is equivalent to minimizing the Log-likelihood of a Bernoulli distributed process using Newton's method. If the problem is written in vector matrix form, with.

parameters $\mathbf{w}^T = [\beta_0, \beta_1, \beta_2, \ldots]$, explanatory variables $\mathbf{x}(i) = [1, x_1(i), x_2(i), \ldots]^T$ and expected value of the Bernoulli distribution $\mu(i) = \frac{1}{1+e^{-\mathbf{w}^T\mathbf{x}(i)}}$, the parameters $\mathbf{w}$ can be found using the following iterative algorithm:

$$w_{k+1} = \left(X^T S_k X\right)^{-1} X^T \left(S_k X w_k + y - \mu_k\right)$$

where $\mathbf{S} = \text{diag}(\mu(i)(1-\mu(i)))$ is a diagonal weighting matrix, $\mu = [\mu(1), \mu(2), \ldots]$ the vector of expected values,

$$\mathbf{X} = \begin{bmatrix} 1 & x_1(1) & x_2(1) & \ldots \\ 1 & x_1(2) & x_2(2) & \ldots \\ \vdots & \vdots & \vdots & \end{bmatrix}$$

the regressor matrix and $\mathbf{y}(i) = [y(1), y(2), \ldots]^T$ the vector of response variables. More details can be found e.g. here.

Evaluating Goodness of Fit

Discrimination in linear regression models is generally measured using R^2. Since this has no direct analog in logistic regression, various methods including the following can be used instead.

Deviance and Likelihood Ratio Tests

In linear regression analysis, one is concerned with partitioning variance via the sum of squares calculations – variance in the criterion is essentially divided into variance accounted for by the predictors and residual variance. In logistic regression analysis, deviance is used in lieu of sum of squares calculations. Deviance is analogous to the sum of squares calculations in linear regression and is a measure of the lack of fit to the data in a logistic regression model. When a "saturated" model is available (a model with a theoretically perfect fit), deviance is calculated by comparing a given model with the saturated model. This computation gives the likelihood-ratio test:

$$D = -2\ln\frac{\text{likelihood of the fitted model}}{\text{likelihood of the saturated model}}.$$

In the above equation D represents the deviance and ln represents the natural logarithm. The log of this likelihood ratio (the ratio of the fitted model to the saturated model) will produce a negative value, hence the need for a negative sign. D can be shown to follow an approximate chi-squared distribution. Smaller values indicate better fit as the fitted model deviates less from the saturated model. When assessed upon a chi-square distribution, nonsignificant chi-square values indicate very little unexplained variance and thus, good model fit. Conversely, a significant chi-square value indicates that a significant amount of the variance is unexplained.

When the saturated model is not available (a common case), deviance is calculated simply as −2·(log likelihood of the fitted model), and the reference to the saturated model's log likelihood can be removed from all that follows without harm.

Two measures of deviance are particularly important in logistic regression: null deviance and model deviance. The null deviance represents the difference between a model with only the intercept (which means "no predictors") and the saturated model. The model deviance represents the difference between a model with at least one predictor and the saturated model. In this respect, the null model provides a baseline upon which to compare predictor models. Given that deviance is a measure of the difference between a given model and the saturated model, smaller values indicate better fit. Thus, to assess the contribution of a predictor or set of predictors, one can subtract the model deviance from the null deviance and assess the difference on a χ^2_{s-p}, chi-square distribution with degrees of freedom equal to the difference in the number of parameters estimated.

Let

$$D_{\text{null}} = -2\ln\frac{\text{likelihood of null model}}{\text{likelihood of the saturated model}}$$

$$D_{\text{fitted}} = -2\ln\frac{\text{likelihood of fitted model}}{\text{likelihood of the saturated model}}.$$

Then the difference of both is:

$$D_{null} - D_{fitted} = -2\left(\ln \frac{\text{likelihood of null model}}{\text{likelihood of the saturated model}} - \ln \frac{\text{likelihood of fitted model}}{\text{likelihood of the saturated model}}\right)$$

$$= -2\ln \frac{\left(\dfrac{\text{likelihood of null model}}{\text{likelihood of the saturated model}}\right)}{\left(\dfrac{\text{likelihood of fitted model}}{\text{likelihood of the saturated model}}\right)}$$

$$= -2\ln \frac{\text{likelihood of the null model}}{\text{likelihood of fitted model}}.$$

If the model deviance is significantly smaller than the null deviance then one can conclude that the predictor or set of predictors significantly improved model fit. This is analogous to the *F*-test used in linear regression analysis to assess the significance of prediction.

Pseudo-R^2s

In linear regression the squared multiple correlation, R^2 is used to assess goodness of fit as it represents the proportion of variance in the criterion that is explained by the predictors. In logistic regression analysis, there is no agreed upon analogous measure, but there are several competing measures each with limitations.

Four of the most commonly used indices and one less commonly used one are examined on this page:

- Likelihood ratio R^2_L.
- Cox and Snell R^2_{CS}.
- Nagelkerke R^2_N.
- McFadden R^2_{McF}.
- Tjur R^2_T.

R^2_L is given by:

$$R_L^2 = \frac{D_{null} - D_{fitted}}{D_{null}}.$$

This is the most analogous index to the squared multiple correlation in linear regression. It represents the proportional reduction in the deviance wherein the deviance is treated as a measure of variation analogous but not identical to the variance in linear regression analysis. One limitation of the likelihood ratio R^2 is that it is not monotonically related to the odds ratio, meaning that it does not necessarily increase as the odds ratio increases and does not necessarily decrease as the odds ratio decreases.

R^2_{CS} is an alternative index of goodness of fit related to the R^2 value from linear regression. It is given by:

$$R^2_{\text{CS}} = 1 - \left(\frac{L_M}{L_0}\right)^{2/n}.$$

where L_M and L_0 are the likelihoods for the model being fitted and the null model, respectively. The Cox and Snell index is problematic as its maximum value is $1 - L_0^{2/n}$. The highest this upper bound can be is 0.75, but it can easily be as low as 0.48 when the marginal proportion of cases is small.

R^2_N provides a correction to the Cox and Snell R^2 so that the maximum value is equal to 1. Nevertheless, the Cox and Snell and likelihood ratio R^2s show greater agreement with each other than either does with the Nagelkerke R^2. Of course, this might not be the case for values exceeding .75 as the Cox and Snell index is capped at this value. The likelihood ratio R^2 is often preferred to the alternatives as it is most analogous to R^2 in linear regression, is independent of the base rate (both Cox and Snell and Nagelkerke R^2s increase as the proportion of cases increase from 0 to .5) and varies between 0 and 1.

R^2_{McF} is defined as

$$R^2_{\text{McF}} = 1 - \frac{\ln(L_M)}{\ln(L_0)},$$

and is preferred over R^2_{CS} by Allison. The two expressions R^2_{McF} and R^2_{CS} are then related respectively by,

$$R^2_{\text{CS}} = 1 - \left(\frac{1}{L_0}\right)^{\frac{2(R^2_{\text{McF}})}{n}}$$

$$R^2_{\text{McF}} = -\frac{n}{2} \cdot \frac{\ln(1 - R^2_{\text{CS}})}{\ln(L_0)}$$

However, Allison now prefers R^2_T which is a relatively new measure developed by Tjur. It can be calculated in two steps:

1. For each level of the dependent variable, find the mean of the predicted probabilities of an event.
2. Take the absolute value of the difference between these means.

A word of caution is in order when interpreting pseudo-R^2 statistics. The reason these indices of fit are referred to as *pseudo* R^2 is that they do not represent the proportionate reduction in error as the

R^2 in linear regression does. Linear regression assumes homoscedasticity, that the error variance is the same for all values of the criterion. Logistic regression will always be heteroscedastic – the error variances differ for each value of the predicted score. For each value of the predicted score there would be a different value of the proportionate reduction in error. Therefore, it is inappropriate to think of R^2 as a proportionate reduction in error in a universal sense in logistic regression.

Hosmer–Lemeshow Test

The Hosmer–Lemeshow test uses a test statistic that asymptotically follows a χ^2 distribution to assess whether or not the observed event rates match expected event rates in subgroups of the model population. This test is considered to be obsolete by some statisticians because of its dependence on arbitrary binning of predicted probabilities and relative low power.

Coefficients

After fitting the model, it is likely that researchers will want to examine the contribution of individual predictors. To do so, they will want to examine the regression coefficients. In linear regression, the regression coefficients represent the change in the criterion for each unit change in the predictor. In logistic regression, however, the regression coefficients represent the change in the logit for each unit change in the predictor. Given that the logit is not intuitive, researchers are likely to focus on a predictor's effect on the exponential function of the regression coefficient – the odds ratio. In linear regression, the significance of a regression coefficient is assessed by computing a *t* test. In logistic regression, there are several different tests designed to assess the significance of an individual predictor, most notably the likelihood ratio test and the Wald statistic.

Likelihood Ratio Test

The likelihood-ratio test discussed above to assess model fit is also the recommended procedure to assess the contribution of individual "predictors" to a given model. In the case of a single predictor model, one simply compares the deviance of the predictor model with that of the null model on a chi-square distribution with a single degree of freedom. If the predictor model has a significantly smaller deviance (c.f chi-square using the difference in degrees of freedom of the two models), then one can conclude that there is a significant association between the "predictor" and the outcome. Although some common statistical packages (e.g. SPSS) do provide likelihood ratio test statistics, without this computationally intensive test it would be more difficult to assess the contribution of individual predictors in the multiple logistic regression case. To assess the contribution of individual predictors one can enter the predictors hierarchically, comparing each new model with the previous to determine the contribution of each predictor. There is some debate among statisticians about the appropriateness of so-called "stepwise" procedures. The fear is that they may not preserve nominal statistical properties and may become misleading.

Wald Statistic

Alternatively, when assessing the contribution of individual predictors in a given model, one may examine the significance of the Wald statistic. The Wald statistic, analogous to the *t*-test in linear regression, is used to assess the significance of coefficients. The Wald statistic is the ratio of the square of the regression coefficient to the square of the standard error of the coefficient and is asymptotically distributed as a chi-square distribution.

$$W_j = \frac{\beta_j^2}{SE_{\beta_j}^2}$$

Although several statistical packages (e.g., SPSS, SAS) report the Wald statistic to assess the contribution of individual predictors, the Wald statistic has limitations. When the regression coefficient is large, the standard error of the regression coefficient also tends to be large increasing the probability of Type-II error. The Wald statistic also tends to be biased when data are sparse.

Case-control sampling

Suppose cases are rare. Then we might wish to sample them more frequently than their prevalence in the population. For example, suppose there is a disease that affects 1 person in 10,000 and to collect our data we need to do a complete physical. It may be too expensive to do thousands of physicals of healthy people in order to obtain data for only a few diseased individuals. Thus, we may evaluate more diseased individuals, perhaps all of the rare outcomes. This is also retrospective sampling, or equivalently it is called unbalanced data. As a rule of thumb, sampling controls at a rate of five times the number of cases will produce sufficient control data.

Logistic regression is unique in that it may be estimated on unbalanced data, rather than randomly sampled data, and still yield correct coefficient estimates of the effects of each independent variable on the outcome. That is to say, if we form a logistic model from such data, if the model is correct in the general population, the β_j parameters are all correct except for β_0. We can correct β_0 if we know the true prevalence as follows:

$$\widehat{\beta_0^*} = \widehat{\beta_0} + \log\frac{\pi}{1-\pi} - \log\frac{\tilde{\pi}}{1-\tilde{\pi}}$$

where π is the true prevalence and $\tilde{\pi}$ is the prevalence in the sample.

Formal Mathematical Specification

There are various equivalent specifications of logistic regression, which fit into different types of more general models. These different specifications allow for different sorts of useful generalizations.

Setup

The basic setup of logistic regression is the same as for standard linear regression.

It is assumed that we have a series of N observed data points. Each data point i consists of a set of m explanatory variables $x_{1,i} \ldots x_{m,i}$ (also called independent variables, predictor variables, input variables, features, or attributes), and an associated binary-valued outcome variable Y_i (also known as a dependent variable, response variable, output variable, outcome variable or class variable), i.e. it can assume only the two possible values 0 (often meaning "no" or "failure") or 1 (often meaning "yes" or "success"). The goal of logistic regression is to explain the relationship between the explanatory variables and the outcome, so that an outcome can be predicted for a new set of explanatory variables.

Some examples:

- The observed outcomes are the presence or absence of a given disease (e.g. diabetes) in a set of patients, and the explanatory variables might be characteristics of the patients thought to be pertinent (sex, race, age, blood pressure, body-mass index, etc.).
- The observed outcomes are the votes (e.g. Democratic or Republican) of a set of people in an election, and the explanatory variables are the demographic characteristics of each person (e.g. sex, race, age, income, etc.). In such a case, one of the two outcomes is arbitrarily coded as 1, and the other as 0.

As in linear regression, the outcome variables Y_i are assumed to depend on the explanatory variables $x_{1,i} \dots x_{m,i}$.

Explanatory Variables

As shown above in the above examples, the explanatory variables may be of any type: real-valued, binary, categorical, etc. The main distinction is between continuous variables (such as income, age and blood pressure) and discrete variables (such as sex or race). Discrete variables referring to more than two possible choices are typically coded using dummy variables (or indicator variables), that is, separate explanatory variables taking the value 0 or 1 are created for each possible value of the discrete variable, with a 1 meaning "variable does have the given value" and a 0 meaning "variable does not have that value". For example, a four-way discrete variable of blood type with the possible values "A, B, AB, O" can be converted to four separate two-way dummy variables, "is-A, is-B, is-AB, is-O", where only one of them has the value 1 and all the rest have the value 0. This allows for separate regression coefficients to be matched for each possible value of the discrete variable. (In a case like this, only three of the four dummy variables are independent of each other, in the sense that once the values of three of the variables are known, the fourth is automatically determined. Thus, it is necessary to encode only three of the four possibilities as dummy variables. This also means that when all four possibilities are encoded, the overall model is not identifiable in the absence of additional constraints such as a regularization constraint. Theoretically, this could cause problems, but in reality almost all logistic regression models are fitted with regularization constraints.)

Outcome Variables

Formally, the outcomes Y_i are described as being Bernoulli-distributed data, where each outcome is determined by an unobserved probability p_i that is specific to the outcome at hand, but related to the explanatory variables. This can be expressed in any of the following equivalent forms:

$$Y_i \mid x_{1,i}, \dots, x_{m,i} \sim \text{Bernoulli}(p_i)$$

$$\mathbb{E}[Y_i \mid x_{1,i}, \dots, x_{m,i}] = p_i$$

$$\Pr(Y_i = y \mid x_{1,i}, \dots, x_{m,i}) = \begin{cases} p_i & \text{if } y = 1 \\ 1 - p_i & \text{if } y = 0 \end{cases}$$

$$\Pr(Y_i = y \mid x_{1,i}, \dots, x_{m,i}) = p_i^y (1 - p_i)^{(1-y)}$$

The meanings of these four lines are:

1. The first line expresses the probability distribution of each Y_i: Conditioned on the explanatory variables, it follows a Bernoulli distribution with parameters p_i, the probability of the outcome of 1 for trial *i*. As noted above, each separate trial has its own probability of success, just as each trial has its own explanatory variables. The probability of success p_i is not observed, only the outcome of an individual Bernoulli trial using that probability.
2. The second line expresses the fact that the expected value of each Y_i is equal to the probability of success p_i, which is a general property of the Bernoulli distribution. In other words, if we run a large number of Bernoulli trials using the same probability of success p_i, then take the average of all the 1 and 0 outcomes, then the result would be close to p_i. This is because doing an average this way simply computes the proportion of successes seen, which we expect to converge to the underlying probability of success.
3. The third line writes out the probability mass function of the Bernoulli distribution, specifying the probability of seeing each of the two possible outcomes.
4. The fourth line is another way of writing the probability mass function, which avoids having to write separate cases and is more convenient for certain types of calculations. This relies on the fact that Y_i can take only the value 0 or 1. In each case, one of the exponents will be 1, "choosing" the value under it, while the other is 0, "canceling out" the value under it. Hence, the outcome is either p_i or $1 - p_i$, as in the previous line.

Linear predictor function

The basic idea of logistic regression is to use the mechanism already developed for linear regression by modeling the probability p_i using a linear predictor function, i.e. a linear combination of the explanatory variables and a set of regression coefficients that are specific to the model at hand but the same for all trials. The linear predictor function $f(i)$ for a particular data point *i* is written as:

$$f(i) = \beta_0 + \beta_1 x_{1,i} + \cdots + \beta_m x_{m,i},$$

where $\beta_0, \ldots, \beta_m$ are regression coefficients indicating the relative effect of a particular explanatory variable on the outcome.

The model is usually put into a more compact form as follows:

- The regression coefficients $\beta_0, \beta_1, \ldots, \beta_m$ are grouped into a single vector β of size $m + 1$.
- For each data point *i*, an additional explanatory pseudo-variable $x_{0,i}$ is added, with a fixed value of 1, corresponding to the intercept coefficient β_0.
- The resulting explanatory variables $x_{0,i}, x_{1,i}, \ldots, x_{m,i}$ are then grouped into a single vector $\boldsymbol{X}_i$ of size $m + 1$.

This makes it possible to write the linear predictor function as follows:

$$f(i) = \beta \cdot X_i,$$

using the notation for a dot product between two vectors.

As a Generalized Linear Model

The particular model used by logistic regression, which distinguishes it from standard linear regression and from other types of regression analysis used for binary-valued outcomes, is the way the probability of a particular outcome is linked to the linear predictor function:

$$\operatorname{logit}(\mathbb{E}[Y_i \mid x_{1,i},\ldots,x_{m,i}]) = \operatorname{logit}(p_i) = \ln\left(\frac{p_i}{1-p_i}\right) = \beta_0 + \beta_1 x_{1,i} + \cdots + \beta_m x_{m,i}$$

Written using the more compact notation described above, this is:

$$\operatorname{logit}(\mathbb{E}[Y_i \mid \mathrm{X}_i]) = \operatorname{logit}(p_i) = \ln\left(\frac{p_i}{1-p_i}\right) = \beta \cdot X_i$$

This formulation expresses logistic regression as a type of generalized linear model, which predicts variables with various types of probability distributions by fitting a linear predictor function of the above form to some sort of arbitrary transformation of the expected value of the variable.

The intuition for transforming using the logit function (the natural log of the odds) was explained above. It also has the practical effect of converting the probability (which is bounded to be between 0 and 1) to a variable that ranges over $(-\infty,+\infty)$ — thereby matching the potential range of the linear prediction function on the right side of the equation.

Note that both the probabilities p_i and the regression coefficients are unobserved, and the means of determining them is not part of the model itself. They are typically determined by some sort of optimization procedure, e.g. maximum likelihood estimation, that finds values that best fit the observed data (i.e. that give the most accurate predictions for the data already observed), usually subject to regularization conditions that seek to exclude unlikely values, e.g. extremely large values for any of the regression coefficients. The use of a regularization condition is equivalent to doing maximum a posteriori (MAP) estimation, an extension of maximum likelihood. (Regularization is most commonly done using a squared regularizing function, which is equivalent to placing a zero-mean Gaussian prior distribution on the coefficients, but other regularizers are also possible.) Whether or not regularization is used, it is usually not possible to find a closed-form solution; instead, an iterative numerical method must be used, such as iteratively reweighted least squares (IRLS) or, more commonly these days, a quasi-Newton method such as the L-BFGS method.

The interpretation of the β_j parameter estimates is as the additive effect on the log of the odds for a unit change in the *j*th explanatory variable. In the case of a dichotomous explanatory variable, for instance gender, e^{β} is the estimate of the odds of having the outcome for, say, males compared with females.

An equivalent formula uses the inverse of the logit function, which is the logistic function, i.e.:

$$\mathbb{E}[Y_i \mid \mathrm{X}_i] = p_i = \operatorname{logit}^{-1}(\beta \cdot X_i) = \frac{1}{1+e^{-\beta \cdot X_i}}$$

The formula can also be written as a probability distribution (specifically, using a probability mass function):

$$\Pr(Y_i = y \mid \mathbf{X}_i) = p_i^{\,y}(1-p_i)^{1-y} = \left(\frac{e^{\beta\cdot X_i}}{1+e^{\beta\cdot X_i}}\right)^{y}\left(1-\frac{e^{\beta\cdot X_i}}{1+e^{\beta\cdot X_i}}\right)^{1-y} = \frac{e^{\beta\cdot X_i\cdot y}}{1+e^{\beta\cdot X_i}}$$

As a Latent-variable Model

The above model has an equivalent formulation as a latent-variable model. This formulation is common in the theory of discrete choice models, and makes it easier to extend to certain more complicated models with multiple, correlated choices, as well as to compare logistic regression to the closely related probit model.

Imagine that, for each trial i, there is a continuous latent variable Y_i^* (i.e. an unobserved random variable) that is distributed as follows:

$$Y_i^* = \beta \cdot X_i + \varepsilon$$

where,

$$\varepsilon \sim \text{Logistic}(0,1)$$

i.e. the latent variable can be written directly in terms of the linear predictor function and an additive random error variable that is distributed according to a standard logistic distribution.

Then Y_i can be viewed as an indicator for whether this latent variable is positive:

$$Y_i = \begin{cases} 1 & \text{if } Y_i^* > 0 \;\text{ i.e. } -\varepsilon < \beta \cdot X_i, \\ 0 & \text{otherwise.} \end{cases}$$

The choice of modeling the error variable specifically with a standard logistic distribution, rather than a general logistic distribution with the location and scale set to arbitrary values, seems restrictive, but in fact it is not. It must be kept in mind that we can choose the regression coefficients ourselves, and very often can use them to offset changes in the parameters of the error variable's distribution. For example, a logistic error-variable distribution with a non-zero location parameter μ (which sets the mean) is equivalent to a distribution with a zero location parameter, where μ has been added to the intercept coefficient. Both situations produce the same value for Y_i^* regardless of settings of explanatory variables. Similarly, an arbitrary scale parameter s is equivalent to setting the scale parameter to 1 and then dividing all regression coefficients by s. In the latter case, the resulting value of Y_i^* will be smaller by a factor of s than in the former case, for all sets of explanatory variables — but critically, it will always remain on the same side of 0, and hence lead to the same Y_i choice.

(Note that this predicts that the irrelevancy of the scale parameter may not carry over into more complex models where more than two choices are available.)

It turns out that this formulation is exactly equivalent to the preceding one, phrased in terms of the generalized linear model and without any latent variables. This can be shown as follows, using the fact that the cumulative distribution function (CDF) of the standard logistic distribution is the logistic function, which is the inverse of the logit function, i.e.

$$\Pr(\varepsilon < x) = \text{logit}^{-1}(x)$$

Then:

$$\begin{aligned}
\Pr(Y_i = 1 \mid X_i) &= \Pr(Y_i^* > 0 \mid X_i) \\
&= \Pr(\beta \cdot X_i + \varepsilon > 0) \\
&= \Pr(\varepsilon > -\beta \cdot X_i) \\
&= \Pr(\varepsilon < \beta \cdot X_i) && \text{(becouse the logistics distribution is symmetric)} \\
&= \text{logit}^{-1}(\beta \cdot X_i) \\
&= p_i
\end{aligned}$$

This formulation—which is standard in discrete choice models—makes clear the relationship between logistic regression (the "logit model") and the probit model, which uses an error variable distributed according to a standard normal distribution instead of a standard logistic distribution. Both the logistic and normal distributions are symmetric with a basic unimodal, "bell curve" shape. The only difference is that the logistic distribution has somewhat heavier tails, which means that it is less sensitive to outlying data (and hence somewhat more robust to model mis-specifications or erroneous data).

Two-way latent-variable Model

Yet another formulation uses two separate latent variables:

$$\begin{aligned}
Y_i^{0*} &= \beta_0 \cdot X_i + \varepsilon_0 \\
Y_i^{1*} &= \beta_1 \cdot X_i + \varepsilon_1
\end{aligned}$$

where

$$\begin{aligned}
\varepsilon_0 &\sim \text{EV}_1(0,1) \\
\varepsilon_1 &\sim \text{EV}_1(0,1)
\end{aligned}$$

where $EV_1(0,1)$ is a standard type-1 extreme value distribution: i.e.

$$\Pr(\varepsilon_0 = x) = \Pr(\varepsilon_1 = x) = e^{-x}e^{-e^{-x}}$$

Then

$$Y_i = \begin{cases} 1 & \text{if } Y_i^{1*} > Y_i^{0*}, \\ 0 & \text{otherwise.} \end{cases}$$

This model has a separate latent variable and a separate set of regression coefficients for each possible outcome of the dependent variable. The reason for this separation is that it makes it easy to extend logistic regression to multi-outcome categorical variables, as in the multinomial logit model. In such a model, it is natural to model each possible outcome using a different set of regression coefficients. It is also possible to motivate each of the separate latent variables as the theoretical utility associated with making the associated choice, and thus motivate logistic regression in terms of utility theory. (In terms of utility theory, a rational actor always chooses the choice with the greatest associated utility.) This is the approach taken by economists when formulating discrete choice models, because it both provides a theoretically strong foundation and facilitates intuitions about the model, which in turn makes it easy to consider various sorts of extensions.

The choice of the type-1 extreme value distribution seems fairly arbitrary, but it makes the mathematics work out, and it may be possible to justify its use through rational choice theory.

It turns out that this model is equivalent to the previous model, although this seems non-obvious, since there are now two sets of regression coefficients and error variables, and the error variables have a different distribution. In fact, this model reduces directly to the previous one with the following substitutions:

$$\beta = \beta_1 - \beta_0$$

$$\varepsilon = \varepsilon_1 - \varepsilon_0$$

An intuition for this comes from the fact that, since we choose based on the maximum of two values, only their difference matters, not the exact values — and this effectively removes one degree of freedom. Another critical fact is that the difference of two type-1 extreme-value-distributed variables is a logistic distribution, i.e. $\varepsilon = \varepsilon_1 - \varepsilon_0 \sim \text{Logistic}(0,1)$. We can demonstrate the equivalent as follows:

$$\begin{aligned}
\Pr(Y_i = 1 \mid X_i) &= \Pr\left(Y_i^{1*} > Y_i^{0*} \mid X_i\right) \\
&= \Pr\left(Y_i^{1*} - Y_i^{0*} > 0 \mid X_i\right) \\
&= \Pr\left(\beta_1 \cdot X_i + \varepsilon_1 - \left(\beta_0 \cdot X_i + \varepsilon_0\right) > 0\right) \\
&= \Pr\left((\beta_1 \cdot X_i - \beta_0 \cdot X_i) + (\varepsilon_1 - \varepsilon_0) > 0\right) \\
&= \Pr((\beta_1 - \beta_0) \cdot X_i + (\varepsilon_1 - \varepsilon_0) > 0) \\
&= \Pr((\beta_1 - \beta_0) \cdot X_i + \varepsilon > 0) && \text{(substitute } \varepsilon \text{ as above)} \\
&= \Pr(\beta \cdot X_i + \varepsilon > 0) && \text{(substitute } \beta \text{ as above)} \\
&= \Pr(\varepsilon > -\beta \cdot X_i) && \text{(now, same as above model)} \\
&= \Pr(\varepsilon < \beta \cdot X_i) \\
&= \text{logit}^{-1}(\beta \cdot X_i) \\
&= p_i
\end{aligned}$$

Example

As an example, consider a province-level election where the choice is between a right-of-center party, a left-of-center party, and a secessionist party (e.g. the Parti Québécois, which wants Quebec to secede from Canada). We would then use three latent variables, one for each choice. Then, in accordance with utility theory, we can then interpret the latent variables as expressing the utility that results from making each of the choices. We can also interpret the regression coefficients as indicating the strength that the associated factor (i.e. explanatory variable) has in contributing to the utility – or more correctly, the amount by which a unit change in an explanatory variable changes the utility of a given choice. A voter might expect that the right-of-center party would lower taxes, especially on rich people. This would give low-income people no benefit, i.e. no change in utility (since they usually don't pay taxes); would cause moderate benefit (i.e. somewhat more money, or moderate utility increase) for middle-incoming people; and would cause significant benefits for high-income people. On the other hand, the left-of-center party might be expected to raise taxes and offset it with increased welfare and other assistance for the lower and middle classes. This would cause significant positive benefit to low-income people, perhaps weak benefit to middle-income people, and significant negative benefit to high-income people. Finally, the secessionist party would take no direct actions on the economy, but simply secede. A low-income or middle-income voter might expect basically no clear utility gain or loss from this, but a high-income voter might expect negative utility, since he/she is likely to own companies, which will have a harder time doing business in such an environment and probably lose money.

These intuitions can be expressed as follows:

Estimated strength of regression coefficient for different outcomes (party choices) and different values of explanatory variables			
	Center-right	**Center-left**	**Secessionist**
High-income	strong +	strong –	strong –
Middle-income	moderate +	weak +	none
Low-income	none	strong +	none

This clearly shows that

1. Separate sets of regression coefficients need to exist for each choice. When phrased in terms of utility, this can be seen very easily. Different choices have different effects on net utility; furthermore, the effects vary in complex ways that depend on the characteristics of each individual, so there need to be separate sets of coefficients for each characteristic, not simply a single extra per-choice characteristic.

2. Even though income is a continuous variable, its effect on utility is too complex for it to be treated as a single variable. Either it needs to be directly split up into ranges, or higher powers of income need to be added so that polynomial regression on income is effectively done.

As a "Log-linear" Model

Yet another formulation combines the two-way latent variable formulation above with the original formulation higher up without latent variables, and in the process provides a link to one of the standard formulations of the multinomial logit.

Here, instead of writing the logit of the probabilities p_i as a linear predictor, we separate the linear predictor into two, one for each of the two outcomes:

$$\ln \Pr(Y_i = 0) = \beta_0 \cdot X_i - \ln Z$$
$$\ln \Pr(Y_i = 1) = \beta_1 \cdot X_i - \ln Z$$

Note that two separate sets of regression coefficients have been introduced, just as in the two-way latent variable model, and the two equations appear a form that writes the logarithm of the associated probability as a linear predictor, with an extra term *-inZ* at the end. This term, as it turns out, serves as the normalizing factor ensuring that the result is a distribution. This can be seen by exponentiating both sides:

$$\Pr(Y_i = 0) = \frac{1}{Z} e^{\beta_0 \cdot \mathbf{X}_i}$$

$$\Pr(Y_i = 1) = \frac{1}{Z} e^{\beta_1 \cdot \mathbf{X}_i}$$

In this form it is clear that the purpose of Z is to ensure that the resulting distribution over Y_i is in fact a probability distribution, i.e. it sums to 1. This means that Z is simply the sum of all un-normalized probabilities, and by dividing each probability by Z, the probabilities become "normalized". That is:

$$Z = e^{\beta_0 \cdot X_i} + e^{\beta_1 \cdot X_i}$$

and the resulting equations are:

$$\Pr(Y_i = 0) = \frac{e^{\beta_0 \cdot X_i}}{e^{\beta_0 \cdot X_i} + e^{\beta_1 \cdot X_i}}$$

$$\Pr(Y_i = 1) = \frac{e^{\beta_1 \cdot X_i}}{e^{\beta_0 \cdot X_i} + e^{\beta_1 \cdot X_i}}.$$

Or generally:

$$\Pr(Y_i = c) = \frac{e^{\beta_c \cdot X_i}}{\sum_h e^{\beta_h \cdot X_i}}$$

This shows clearly how to generalize this formulation to more than two outcomes, as in multinomial logit. Note that this general formulation is exactly the Softmax function as in

$$\Pr(Y_i = c) = \operatorname{softmax}(c, \beta_0 \cdot X_i, \beta_1 \cdot X_i, \ldots).$$

In order to prove that this is equivalent to the previous model, note that the above model is overspecified, in that $\Pr(Y_i = 0)$ and $\Pr(Y_i = 1)$ cannot be independently specified: rather $\Pr(Y_i = 0) + \Pr(Y_i = 1) = 1$ so knowing one automatically determines the other. As a result, the model is nonidentifiable, in that multiple combinations of $\boldsymbol{\beta}_0$ and $\boldsymbol{\beta}_1$ will produce the same probabilities for all possible explanatory variables. In fact, it can be seen that adding any constant vector to both of them will produce the same probabilities:

$$\begin{aligned}\Pr(Y_i = 1) &= \frac{e^{(\beta_1 + \mathbf{C}) \cdot \mathbf{X}_i}}{e^{(\beta_0 + \mathbf{C}) \cdot \mathbf{X}_i} + e^{(\beta_1 + \mathbf{C}) \cdot \mathbf{X}_i}} \\ &= \frac{e^{\beta_1 \cdot \mathbf{X}_i} e^{\mathbf{C} \cdot \mathbf{X}_i}}{e^{\beta_0 \cdot \mathbf{X}_i} e^{\mathbf{C} \cdot \mathbf{X}_i} + e^{\beta_1 \cdot \mathbf{X}_i} e^{\mathbf{C} \cdot \mathbf{X}_i}}\end{aligned}$$

$$= \frac{e^{\mathbf{C}\cdot\mathbf{X}_i} e^{\beta_1\cdot\mathbf{X}_i}}{e^{\mathbf{C}\cdot\mathbf{X}_i}(e^{\beta_0\cdot\mathbf{X}_i} + e^{\beta_1\cdot\mathbf{X}_i})}$$

$$= \frac{e^{\beta_1\cdot\mathbf{X}_i}}{e^{\beta_0\cdot\mathbf{X}_i} + e^{\beta_1\cdot\mathbf{X}_i}}.$$

As a result, we can simplify matters, and restore identifiability, by picking an arbitrary value for one of the two vectors. We choose to set $\beta_0 = \mathbf{0}$. Then,

$$e^{\beta_0\cdot\mathbf{X}_i} = e^{\mathbf{0}\cdot\mathbf{X}_i} = 1$$

and so

$$\Pr(Y_i = 1) = \frac{e^{\beta_1\cdot\mathbf{X}_i}}{1 + e^{\beta_1\cdot\mathbf{X}_i}} = \frac{1}{1 + e^{-\beta_1\cdot\mathbf{X}_i}} = p_i$$

which shows that this formulation is indeed equivalent to the previous formulation. (As in the two-way latent variable formulation, any settings where $\beta = \beta_1 - \beta_0$ will produce equivalent results.)

Note that most treatments of the multinomial logit model start out either by extending the "log-linear" formulation presented here or the two-way latent variable formulation presented above, since both clearly show the way that the model could be extended to multi-way outcomes. In general, the presentation with latent variables is more common in econometrics and political science, where discrete choice models and utility theory reign, while the "log-linear" formulation here is more common in computer science, e.g. machine learning and natural language processing.

As a single-layer Perceptron

The model has an equivalent formulation:

$$p_i = \frac{1}{1 + e^{-(\beta_0 + \beta_1 x_{1,i} + \cdots + \beta_k x_{k,i})}}.$$

This functional form is commonly called a single-layer perceptron or single-layer artificial neural network. A single-layer neural network computes a continuous output instead of a step function. The derivative of p_i with respect to $X = (x_1, ..., x_k)$ is computed from the general form:

$$y = \frac{1}{1 + e^{-f(X)}}$$

where $f(X)$ is an analytic function in X. With this choice, the single-layer neural network is identical to the logistic regression model. This function has a continuous derivative, which allows it to be used in backpropagation. This function is also preferred because its derivative is easily calculated:

$$\frac{\mathrm{d}y}{\mathrm{d}X} = y(1 - y)\frac{\mathrm{d}f}{\mathrm{d}X}.$$

In Terms of Binomial Data

A closely related model assumes that each i is associated not with a single Bernoulli trial but with n_i independent identically distributed trials, where the observation Y_i is the number of successes observed (the sum of the individual Bernoulli-distributed random variables), and hence follows a binomial distribution:

$$Y_i \sim \mathrm{Bin}(n_i, p_i), \text{ for } i = 1, \ldots, n$$

An example of this distribution is the fraction of seeds (p_i) that germinate after n_i are planted.

In terms of expected values, this model is expressed as follows:

$$p_i = \mathbb{E}\left[\left.\frac{Y_i}{n_i}\right| \mathbf{X}_i\right],$$

so that:

$$\operatorname{logit}\left(\mathbb{E}\left[\left.\frac{Y_i}{n_i}\right| X_i\right]\right) = \operatorname{logit}(p_i) = \ln\left(\frac{p_i}{1-p_i}\right) = \beta \cdot X_i,$$

Or equivalently:

$$\Pr(Y_i = y \mid \mathbf{X}_i) = \binom{n_i}{y} p_i^y (1-p_i)^{n_i - y} = \binom{n_i}{y}\left(\frac{1}{1+e^{-\beta\cdot\mathbf{X}_i}}\right)^y \left(1 - \frac{1}{1+e^{-\beta\cdot\mathbf{X}_i}}\right)^{n_i - y}.$$

This model can be fit using the same sorts of methods as the above more basic model.

Bayesian

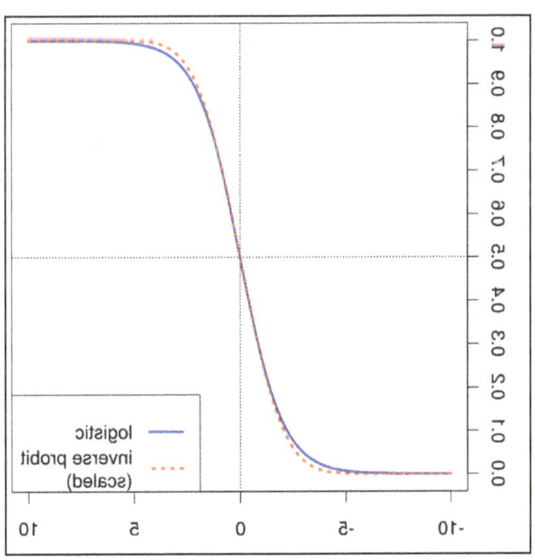

Comparison of logistic function with a scaled inverse probit function (i.e. the CDF of the normal distribution), comparing $\sigma(x)$ vs. $\Phi(\sqrt{\frac{\pi}{8}}x)$, which makes the slopes the same at the origin. This shows the heavier tails of the logistic distribution.

In a Bayesian statistics context, prior distributions are normally placed on the regression coefficients, usually in the form of Gaussian distributions. There is no conjugate prior of the likelihood function in logistic regression. Back when Bayesian inference was performed analytically, this made the posterior distribution difficult to calculate except in very low dimensions. Now, though, automatic software such as OpenBUGS, JAGS and Stan allow these posteriors to be computed using simulation, so lack of conjugacy is not a concern. However, when the sample size or the number of parameters is large, full Bayesian simulation can be slow, and people often use approximate methods such as variational Bayes and expectation propagation.

Estimation of Mortality and Fertility

In 1950's when the studies of population started in both developed and developing countries, lack of data on the vital rates was a big problem for most countries. This was the best period for development of mathematical models in population studies. Mathematicians and statisticians developed models to estimate birth rate, death rate, total fertility rate and life expectancy using incomplete and unreliable data (Shryock et al., 1971; Roy, 1987). The main issue was: can we get working estimates of fertility and mortality using census growth rate, age distribution of population, and open and closed birth intervals found from survey data? Mathematical models helped in this greatly. For example, when reliable data on total fertility rate (TFR) did not exist but data on parity by age was available from surveys, using an empirical relationship, Coale and Demeny suggested the following approximate formula:

$$TFR = \frac{P_3^2}{P_2}$$

Gompertz Model

Brass showed that if the fertility pattern can be described by a Gompertz function of the proportion experienced by each age, then the following is a better estimate of TFR than the above (Brass, 1979):

$$P_2\left(\frac{P_4}{P_3}\right)^4$$

With improvement in data sources and development of new sources such as Sample Registration Scheme, the need for such models declined.

Knud (1983) compared cancer mortality between sexes, cohorts and cities by using Poisson distribution for number of deaths at a particular age and the mortality rate (defined as chance of survival to age x) as follows:

$$l_x = b^{x^k}$$

b and k are two parameters for which maximum likelihood estimates were obtained.

Gompertz model has been commonly applied for studying mortality. Using an age-dependent shape parameter, Weon (2004) used a Weibull model for mortality rate μ(t), i.e., ratio of density (f(t)) and survival functions (S(t) = 1 – F(t)) for estimating maximum longevity, as follows,

$$S(t) = \exp(-t/a)^{\beta(t)}$$

$$\mu(t) = (t/a)^{\beta(t)} * [\frac{\beta(t)}{t} + \ln(t/a) * \frac{d\beta t}{dt}]$$

He assumed that the typical human survival curves show: a rapid decrease in survival in the first few years of life, a relatively steady decrease, and then an abrupt decrease near death.

In the absence of data on measures of fertility, they were derived from surveys. In India, demographers derived parity progression ratios from survey data and used exponential model, displaced exponential model, Poisson model and life table approach for birth intervals (Pathak, 1989; Krishnamoorthy, 1989; Pathak and Ram, 1989).

Gompertz Function

A Gompertz curve or Gompertz function, named after Benjamin Gompertz, is a sigmoid function. It is a type of mathematical model for a time series, where growth is slowest at the start and end of a time period. The right-hand or future value asymptote of the function is approached much more gradually by the curve than the left-hand or lower valued asymptote, in contrast to the simple logistic function in which both asymptotes are approached by the curve symmetrically. It is a special case of the generalised logistic function.

Formula:

$$y(t) = ae^{-be^{-ct}},$$

where

- a is an asymptote, since $\lim_{t\to\infty} ae^{-be^{-ct}} = ae^0 = a$.
- b, c are positive numbers.
- b sets the displacement along the x-axis (translates the graph to the left or right).
- c sets the growth rate (y scaling).
- e is Euler's Number ($e = 2.71828...$).

Derivation

The function curve can be derived from a Gompertz law of mortality, which states the rate of mortality (decay) falls exponentially with current size. Mathematically,

$$k^r \propto \frac{1}{y(t)},$$

where:

- $r = \frac{y'(t)}{y(t)}$ is the rate of growth.
- k is an arbitrary constant.

Example uses

Examples of uses for Gompertz curves include:

- Mobile phone uptake, where costs were initially high (so uptake was slow), followed by a period of rapid growth, followed by a slowing of uptake as saturation was reached.
- Population in a confined space, as birth rates first increase and then slow as resource limits are reached.
- Modeling of growth of tumors.
- Modelling market impact in finance.

Growth of Tumors

In the 1960s A.K. Laird for the first time successfully used the Gompertz curve to fit data of growth of tumors. In fact, tumors are cellular populations growing in a confined space where the availability of nutrients is limited. Denoting the tumor size as X(t) it is useful to write the Gompertz Curve as follows:

$$X(t) = K \exp\left(\log\left(\frac{X(0)}{K}\right)\exp(-\alpha t)\right) = K\left(\frac{X(0)}{K}\right)^{\exp(-\alpha t)}$$

where:

- X(0) is the tumor size at the starting observation time;
- K is the carrying capacity, i.e. the maximum size that can be reached with the available nutrients. In fact it is:

$$\lim_{t \to +\infty} X(t) = K$$

independently on X(0)>0. Note that, in absence of therapies etc.. usually it is X(0)<K, whereas, in presence of therapies, it may be X(0)>K;

- α is a constant related to the proliferative ability of the cells.
- log() refers to the natural log.

It is easy to verify that the dynamics of X(t) is governed by the Gompertz differential equation:

$$X'(t) = \alpha \log\left(\frac{K}{X(t)}\right) X(t)$$

i.e. is of the form:

$$X'(t) = F(X(t))X(t), F'(X) \le 0$$

where F(X) is the instantaneous proliferation rate of the cellular population, whose decreasing nature is due to the competition for the nutrients due to the increase of the cellular population, similarly to the logistic growth rate. However, there is a fundamental difference: in the logistic case the proliferation rate for small cellular population is finite:

$$F(X)=\alpha\left(1-\left(\frac{X}{K}\right)^{\nu}\right)\Rightarrow F(0)=\alpha<+\infty$$

whereas in the Gompertz case the proliferation rate is unbounded:

$$\lim_{X\to 0^+} F(X)=\lim_{X\to 0^+}\alpha\log\left(\frac{K}{X}\right)=+\infty$$

As noticed by Steel and by Wheldon, the proliferation rate of the cellular population is ultimately bounded by the cell division time. Thus, this might be an evidence that the Gompertz equation is not good to model the growth of small tumors. Moreover, more recently it has been noticed that, including the interaction with immune system, Gompertz and other laws characterized by unbounded F(0) would preclude the possibility of immune surveillance.

Gompertz Growth and Logistic Growth

The Gompertz differential equation:

$$X'(t)=\alpha\log\left(\frac{K}{X(t)}\right)X(t)$$

is the limiting case of the generalized logistic differential equation:

$$X'(t)=\alpha\nu\left(1-\left(\frac{X(t)}{K}\right)^{\frac{1}{\nu}}\right)X(t)$$

(where $\nu>0$ is a positive real number) since:

$$\lim_{\nu\to+\infty}\nu\left(1-x^{1/\nu}\right)=-\log(x)$$

In addition, there is an inflection point in the graph of the generalized logistic function when:

$$X(t)=\left(\frac{\nu}{\nu+1}\right)^{\nu}K$$

and one in the graph of the Gompertz function when:

$$X(t)=\frac{K}{e}=K\cdot\lim_{\nu\to+\infty}\left(\frac{\nu}{\nu+1}\right)^{\nu}$$

Gomp-ex Law of Growth

Based on the above considerations, Wheldon proposed a mathematical model of tumor growth, called the Gomp-Ex model, that slightly modifies the Gompertz law. In the Gomp-Ex model it is assumed that initially there is no competition for resources, so that the cellular population expands following the exponential law. However, there is a critical size threshold X_C such that for $X > X_C$ the growth follows the Gompertz Law:

$$F(X) = \max\left(a, \alpha \log\left(\frac{K}{X}\right)\right)$$

so that:

$$X_C = K \exp\left(-\frac{a}{\alpha}\right).$$

Here there are some numerical estimates for X_C:

- $X_C \approx 10^9$ for human tumors.
- $X_C \approx 10^6$ for murine (mouse) tumors.

Prediction and Estimation of Migration

Studies of migration aim at estimation, prediction and explanation of migration. In this context, for the first time Ravenstein published an article in *Journal of the Statistical Society* in 1885 in which he showed that, as believed by William Farr and many others, migration is not without laws. He developed some well known laws of migration, such as follows, which are still difficult to refute:

- The great body of migrants travel short distances.
- Women outnumber men in short distance migration.
- Migrants move from agricultural areas (places of dispersion) to industrial cities (places of absorption) followed by migration from centres of industrial cities to suburban areas and from remote areas to places of dispersion.
- Each migration current has a counter current with similar characteristics.
- The major causes of migration are economic.

Four years later, in 1889, Ravenstein published in the *Journal of Royal Statistical Society*. This article was based on the experience of North America and Europe (based on the experience of UK). In this section, he said that people travel long distances to occupy unsettled land. Nearly 100 years later, in 1940 Samuel A. Stouffer published an article in *American Sociological Review*. He showed that the number of migrants from place i to place j is inversely proportional to intervening opportunities. The model was confirmed by 1n 1975 by Wadycki who found it to be quite an accurate description of migration. Z_{ipf} (1946) developed and validated a gravity model to represent that volume of migration between two cities (M_{ij}) is directly proportion to populations of the cities

(P_i and P_j) and inversely proportional to distance separating the two cities (d_{ij}) He assumed that income and unemployment are uniformly distributed over the areas.

$$M\,ij = K(pi * pj)/D$$

where K is a constant to be found out from the empirical data. To some sociologists the model may look funny or absurd, but Z_{ipf} found that the model fitted very well for all modes of transportation.

After all, if the logic used in the model is sound that accurate data are available, why would mathematical equations not produce a good fit to empirical data?

In 1975, using a similar argument, Dorigo and Tobler (1983) expressed:

$$M_{ij} = K\left(U_i W_i / U_j W_j\right) L_i L_j / d_{ij}$$

Also

$$R_i - E_j = K\left(U_i W_j / U_j W_i\right) L_i L_j$$

where M_{ij} refers to migration from place i to place j. R refers to rejecting, repelling, or repulsing, "push" away factors. E refers to enticing or "pull" toward factors. U, W and L refer to unemployment rate, wage rate and number of people in the labour market.

One can change the variables in the model, delete or add some, transform variables to mathematical or trigonometric form, but as long as the logic developed by Z_{ipf} in suggesting gravity models is correct, the above types of models would remain relevant.

Economic-Demographic Models

Coale and Hoover (1958) pioneered the modelling of demographic and economic variables through which they showed what the short term and long term effects of population growth on economic development are. Their model paved the way for large scale computer simulation models in population and development (Arthur and McNicoll, 1975). These models are developed on the basis of certain axioms, empirical relationships and theoretical understanding of the issues involved. The variables of the system are connected through mathematical equations and the parameters of the equations are estimated using the available data and appropriate estimation techniques. Predictions based on these models are correct to the extent that we have correct assumptions, accurate data, and appropriate estimation procedures. Error at any level may lead to error of an unknown magnitude in the results.

Contemporary Issues in Modelling Population Processes

With improved data on demographic processes, the need for certain types of mathematical models has diminished gradually. Population censuses, improved vital registration systems, new data sources as Sample Registration Scheme in India, demographic health surveys, national surveys such as National Family Health Surveys in India (NFHS), child and reproductive health survey, have provided quite reliable estimates of various demographic measures along with social class and urban-rural differences. Moreover, demographic transition from high to low death and birth rates is occurring in all countries of the world. Now the issues are no more estimation and prediction of demographic rates and ratios. There are other issues: aging and health at the older ages (Crimmins, 2005; Turra, et al., 2005; Lutz and Scherbov, 2005; Bongaarts, 2006; Jansen and Kunst, 2007), timing of the end of world population growth (Lutz et al., 2003, sexuality (Mensch, 2006, Gabrielli, 2007), determinants of neonatal mortality (Vandresse, 2008), and trends in HIV/AIDS (Bongaarts, 2008).

It may be said that models can be used in any field of demographic interest. What one needs to apply modelling is to have:

- Clarity of objectives.
- Appropriate measurement of the variables of the system.
- Theory that links the various variables – directly or indirectly.
- Mathematical equations to describe the above relationships.
- Adequate data of high reliability.
- Adequate computing facilities.
- Instruments to measure the fitness of the model to real world data.

As time passed and demographers took interest in new areas, several new domains of modelling have developed in literature. The area of modelling is going to expand further.

Popularity of Statistical Models Over Mathematical Models

In the recent past, models based on advanced statistical techniques have become more popular than mathematical models (Jansen, 2007; Vandresse, 2008). Poisson regression (expressing log of rates as linear combination of independent variables), logit regression (expressing log odds as linear combination of independent variables) and multiple classification analysis are increasingly used by the social scientists. Guo and Harris (2000) used a structural equation model to examine the effect of poverty and other factors on children's intellectual development.

References

- Islam, Towhidul; Fiebig, Denzil G.; Meade, Nigel (2002), "Modelling multinational telecommunications demand with limited data", International Journal of Forecasting, 18 (4): 605–624, doi:10.1016/S0169-2070(02)00073-0
- Gelman, Andrew; Hill, Jennifer (2007). Data Analysis Using Regression and Multilevel/Hierarchical Models. New York: Cambridge University Press. pp. 79–108. ISBN 978-0-521-68689-1
- Palei, S. K.; Das, S. K. (2009). "Logistic regression model for prediction of roof fall risks in bord and pillar workings in coal mines: An approach". Safety Science. 47: 88–96. doi:10.1016/j.ssci.2008.01.002
- Vittinghoff, E.; McCulloch, C. E. (12 January 2007). "Relaxing the Rule of Ten Events per Variable in Logistic and Cox Regression". American Journal of Epidemiology. 165 (6): 710–718. doi:10.1093/aje/kwk052. PMID 17182981
- Cohen, Jacob; Cohen, Patricia; West, Steven G.; Aiken, Leona S. (2002). Applied Multiple Regression/Correlation Analysis for the Behavioral Sciences (3rd ed.). Routledge. ISBN 978-0-8058-2223-6

Permissions

All chapters in this book are published with permission under the Creative Commons Attribution Share Alike License or equivalent. Every chapter published in this book has been scrutinized by our experts. Their significance has been extensively debated. The topics covered herein carry significant information for a comprehensive understanding. They may even be implemented as practical applications or may be referred to as a beginning point for further studies.

We would like to thank the editorial team for lending their expertise to make the book truly unique. They have played a crucial role in the development of this book. Without their invaluable contributions this book wouldn't have been possible. They have made vital efforts to compile up to date information on the varied aspects of this subject to make this book a valuable addition to the collection of many professionals and students.

This book was conceptualized with the vision of imparting up-to-date and integrated information in this field. To ensure the same, a matchless editorial board was set up. Every individual on the board went through rigorous rounds of assessment to prove their worth. After which they invested a large part of their time researching and compiling the most relevant data for our readers.

The editorial board has been involved in producing this book since its inception. They have spent rigorous hours researching and exploring the diverse topics which have resulted in the successful publishing of this book. They have passed on their knowledge of decades through this book. To expedite this challenging task, the publisher supported the team at every step. A small team of assistant editors was also appointed to further simplify the editing procedure and attain best results for the readers.

Apart from the editorial board, the designing team has also invested a significant amount of their time in understanding the subject and creating the most relevant covers. They scrutinized every image to scout for the most suitable representation of the subject and create an appropriate cover for the book.

The publishing team has been an ardent support to the editorial, designing and production team. Their endless efforts to recruit the best for this project, has resulted in the accomplishment of this book. They are a veteran in the field of academics and their pool of knowledge is as vast as their experience in printing. Their expertise and guidance has proved useful at every step. Their uncompromising quality standards have made this book an exceptional effort. Their encouragement from time to time has been an inspiration for everyone.

The publisher and the editorial board hope that this book will prove to be a valuable piece of knowledge for students, practitioners and scholars across the globe.

Index

www.ingramcontent.com/pod-product-compliance
Lightning Source LLC
LaVergne TN
LVHW080722210926
848485LV00003B/503